Caravan and Motorhome Electrics
Collyn Rivers
RVBooks.com.au (2019)

Caravan and Motorhome Electrics is a technical but accessible guide to all aspects of RV electrics. It explains how things work and what they do (not what vendors claim they do).

Publishing Details

Publisher: RV Books, 2 Scotts Rd, Mitchells Island, NSW, 2430. info@rvbooks.com.au

RV Books: 2019.

Copyright: Collyn Rivers. All rights reserved. Apart from minor extracts for the purposes of review, no part of this publication may be reproduced, stored in a retrieval system, or transmitted in any form or by any means, electronic, mechanical, photocopying, recording, or otherwise, without the written permission of the publisher. Most drawings, graphs and tables in this book are original and copyright.

National Library of Australia - Cataloguing-in-Publication data

Rivers, Collyn

Caravan & Motorhome Electrics

Fourth edition, 2020

ISBN: 978-0-6483190-8-5. Caravan & Motorhome Electrics. Motorhome & Caravan Electrics. I. Title.

Publisher's Note: To ensure topicality this book is updated when necessary.

Disclaimer: Every effort has been made to ensure that the information in this publication is accurate. However no responsibility is accepted by the publisher for any error or omission or for any loss, damage, or injury suffered by anyone relying on the information or advice contained in this publication, or from any other cause. (The author would appreciate feedback relating to any errors and/or omission.)

The author has extensive electrical engineering experience but is not a licensed electrician.

Front Cover Picture: Adapted from 'Abstract colourful wire' by Caravan & Motorhome Books. Copyright (of the original) © Plingho (Dei Ling Hoo) dreamstime.com

Chapter List

A complete table of content is available at the end of this book.

This book's intent	1
Acknowledgments	2
Chapter 1 - Terminology	3
Chapter 2 - Basic Electrics	6
Chapter 3 - Overview of an RV's electrical needs	11
Chapter 4 - Providing the power	18
Chapter 5 - Scaling the power required	23
Chapter 6 - Installing safety and legality	26
Chapter 7 - Installing 12/24 volt wiring	28
Chapter 8 - Mains-voltage wiring	44
Chapter 9 - Batteries (general)	54
Chapter 10 - Installing batteries	64
Chapter 11 - Battery charging (general)	69
Chapter 12 - Battery charging (alternators)	78
Chapter 13 - dc-dc alternator charging	84
Chapter 14 - Installing a dc-dc alternator charger	85
Chapter 15 - Variable voltage alternator charging	86
Chapter 16 - Inverters	88
Chapter 17 - Installing an inverter	93
Chapter 18 - Generators	95
Chapter 19 - Installing a generator	99
Chapter 20 - Wind power generators	103
Chapter 21 - Fuel cells	108
Chapter 22 - Solar energy	113
Chapter 23 - Installing solar modules	122
Chapter 24 - Solar regulators	127
Chapter 25 - Installing solar regulators	129
Chapter 26 - Energy monitoring	131
Chapter 27 - Installing energy monitors	133
Chapter 28 - Lighting	136
Chapter 29 - Water	140
Chapter 30 - Electric toilets	144
Chapter 31 - Refrigerators	146
Chapter 32 - Installing & optimising fridges	153
Chapter 33 - Television	159
Chapter 34 - Communications	164

Chapter 35 - Electrical & radio interference	168
Chapter 36 - Lightning protection	170
Chapter 37 - Caravan specific issues	172
Chapter 38 - Example systems	179
Chapter 39 - Frequently asked questions	185
Chapter 40 - Import electrical issues	187
Chapter 41 - Contacts & references	189
Detailed Table of Contents	191

This book's intent

This book, in its fourth (and now eBook) edition is intended primarily for buyers, owners, designers and builders of camper trailers, holiday and other cabins, caravans, camper vans and motorhomes. It also bridges the still existing gap between the auto-electrical and alternative energy disciplines.

Along the way it provides an insight into major components such as alternators, regulators, batteries and their charging and monitoring, solar energy and motor-generators. It explains why solar modules are rated as they are - and what they really produce. It warns of unrealistic expectations, such as using electric stoves and air-conditioning away from mains power and/or power from large motor-generators.

In essence *Caravan & Motorhome Electrics* shows how to design affordable systems that really do work, how to install them and how to identify, understand and (often) fix anything that does not work as it should.

Caravan & Motorhome Electrics also covers the already happening changes in ways that alternators may be used for auxiliary RV use (particularly battery charging). These changes are primarily intended to reduce energy usage, including by minimising electrical energy drawn by the vehicle's electrical system, with the aim of reducing emissions.

While RV electrical systems are not overly complex, competent installation and service, particularly for the increasing number of RVs with solar, can still be hard to find. *Caravan & Motorhome Electrics*' content therefore includes the installation and rectification of solar systems: this is increasingly necessary, not least as solar technician training does not include vehicle charging systems. Nor do auto-electrical courses cover RV electrics, let alone solar. In most recreational vehicles these systems interact and the general lack of such training leads to many RVs having system problems reported by many owners as being hard to identify and repair.

Partly because of the above, RV owners and auto electricians worldwide bought the previous editions of what was originally *Motorhome Electrics* which right from its 2002 beginnings, covered solar.

The book was totally rewritten and its name changed to *Caravan & Motorhome Electrics* in 2013. The second print run, in September 2014, needed only minor updates. Recent major changes in alternator and battery technology, however, necessitated the publication of a major 2016 revised (second) edition and now further updated with this (2019) eBook edition.

Many readers also buy the associated *Solar That Really Works* (for cabins and RVs), and/or if installing home or property solar, *Solar Success*.

As this book is intended as a working guide, rather than a text book, it has a number of deliberate duplications to ease that intent.

Acknowledgments

Technical proof reading: Ian Brown B.Sc., BE., (the late) Barry Powell, Lawrie Beales and Peter and Margaret Wright. Conversion to eBook, Daniel Weinstein.

Photographs (acknowledged on the pix). Please advise us if any correction is needed.

Chapter 1

Terminology

Any publication dealing with both 12/24 volt dc (direct current) vehicle electrical systems and 230 volt ac (alternating current) electrical systems has difficulty with the electrical term 'low voltage'.

Low voltage, to many without specialised electrical knowledge, tends to be regarded as 12 volts or 24 volts. In electrical engineering, however, Low voltage' (as legally defined by the International Electrotechnical Commission - and in the mutual Australia/New Zealand standard AS/NZS 3000:2018) is 50-1000 volts alternating current (ac) and 120-1500 volts direct current (dc). To an electrical engineer a 230 volt system thus operates at Low voltage.

The term Extra-low voltage applies to any voltage not exceeding 50 volts alternating current, or 120 volts (ripple-free) direct current.

Despite this, even electricians may casually use the term low voltage when discussing 12 volt and 24 volt systems, despite such voltages being legally defined as Extra-low voltage in Australia, New Zealand and many other countries.

Mains voltage in Australia and New Zealand is (legally) 230 volts alternating current. For legal reasons, several sections of this book (relating to 230 volt supply) uses the term Low voltage.

The (initially US) term 'grid' voltage tends to be used increasingly in Australia to imply 'mains-voltage' - but can mislead as the voltage of interstate supply lines in the grid distribution network may be several hundred thousand volts ac!

A few academic readers have criticised the use of the term 'Peak Sun Hour' (PSH), used in this and previous versions of this book, on the basis that it is not academically recognised. Whilst this is so, the term was devised some 50 year ago by the photo-voltaic industry and is used both technically and promotionally by that industry worldwide. There is no choice but to use it in books that are intended for general readership.

Electrical units & terms

Amps: the amount of electrical current that is flowing. It is akin to water flow in a pipe. The greater the voltage, the greater the amount of current that consequently flows. Its common abbreviation is A.

Amp-hour: the amount of electrical current that flows in one hour. A device that generates four amps for five hours thus produces 20 amp hours. Amp hour is commonly abbreviated to Ah.

Amp hours/day: the number of amp hours consumed in a 24 hour period. This unit is handy when scaling solar systems, etc. The correct abbreviation is Ah/day.

Energy: the capacity for doing work. Its base unit is the joule and 559,500 joules/second is 1.0 hp.

Joule: a joule is the work done, or energy expended, when a force of one newton moves the point of application a distance of one metre in the direction of that force. It is also that work done when 1.0 kg is lifted through 0.1 metre. The unit is applicable also to heat: one joule is the amount of energy needed heat 0.001 litre of water by about 0.25°C.

Newton: the force that, when applied to a body having a mass of one kilogram, causes an acceleration of one metre per second in the direction of application of that force.

Ohms: this unit quantifies the resistance to electric current flow. It is either spelled out (i.e. as ohm), or (and traditionally) by the Greek symbol for omega (Ω). One ohm is equal to the resistance of a substance to a flow of one amp if one volt is applied across it.

Ohm's Law: volts, amps and ohms are interrelated and expressed and defined by Ohm's Law. That law states that the direct current (dc) that flows in a circuit is directly proportional to the voltage across that circuit. It is valid for metal circuits and some (but not all) liquids that are electrically conductive.

Power: the rate at which work is done. Its base unit is the watt and is the work done, or expended, at the rate of one joule per second. Power in watts can also be seen as equal to the energy in joules, divided by the time in seconds.

Power factor: the ratio of the average power to the apparent power.

In electrical work generally, numbers over 1000 may use the prefix kilo (the abbreviation is k) for 1000, and mega (the abbreviation is M) for 1,000,000. Hence 1 kW and 1 MW, etc.

Volts: the pressure that causes electricity to flow: akin to pressure in a pipe (abbreviated as V). It is common to indicate whether such voltage is ac or dc - e.g. Vac or Vdc. It is the potential difference between two points on a conductor carrying a constant one amp when the power dissipated between them is equal to one watt.

Resistance: to varying extents substances resist the flow of electricity. This resistance generates heat. Resistance can be useful if heat is required but, where it is not, it wastes energy. The thinner a cable, the more it resists the flow of current. In doing so it heats up, resulting in electrical energy being lost as heat. The term 're sistance' is often abbreviated to R. It is measured in ohms, (see below).

Watts: a watt is a measure of energy used when work is done, or energy used at the rate of one joule per second. Electrically, 1.0 watt is the product of 1.0 amp and 1.0 volt. See also Energy/Power below.

Watt hours: one watt hour is an energy usage of one watt for one hour. It is abbreviated as Wh.

Watt hours/day: as with amp hours/day, watt hours/day are watts per hour over a 24 hour period.

Chapter 2
Basic Electrics

Lord Kelvin - source unknown

Sometime in the mid-1850s, scientific pioneer, Lord Kelvin was lecturing on electricity. He asked his class: "What is electricity". One student put his hand up but then stammered out that he'd forgotten. Lord Kelvin turned slowly to the class and said:

"Gentlemen, you have just witnessed the greatest tragedy of this century. Only two people know what electricity is. One is God, and the other one is Mr Smith. God won't tell us - and Mr Smith has forgotten."

Since that day, any number of theories have attempted to explain it and for some 100 years, settled on an explanation that worked well enough as a model that enables engineers to make calculations and design and build sophisticated equipment. The reality is, however, that to this day, electricity's exact nature has yet to be fully understood.

The model mostly used (Figure 1.2) assumes a universe formed of atoms in structures called molecules. Each molecule has a nucleus of one or more protons and neutrons, in effect, forming a single unit. Around each nucleus, electrons whirl at vast speed.

The resultant force attempting to hurl electrons out of orbit is counteracted by an attractive force that maintains the whirling electrons in orbit around the protons and neutrons.

In materials like metal, some electrons (given an incentive) flow around a conductive loop - such as copper wire. That incentive may be chemical (a battery), sunlight on a form of silicon (solar), wire moving in a magnetic field (an alternator), or squeezing quartz (discovered by Volta in 1780). The 'force' of electron flow, called 'voltage,' is akin to water pressure in a pipe.

In some materials, most electrons are rigidly bound to the central nucleus and unable to flow. Such materials are called 'insulators'. Most resist electron flow but do not totally prevent it.

Electron flow

The number of electrons that flow is huge: in a small torch it is about 10,000,000,000,000,000,000 electrons (10^{18}) a second. The unit used, (the amp) is about 6.24×10^{18} electrons/second. Electron flow starts and stops at once but individual electrons only move a few centimetres a minute.

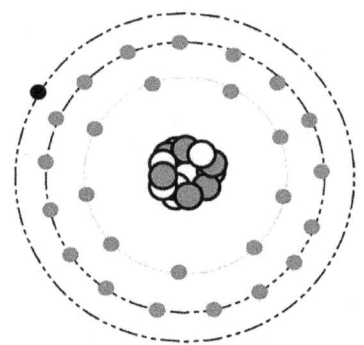

Figure 1.2. In this traditional representation, an atom's nucleus of bound protons and neutrons are circled by orbiting electrons. If stripped off, as the one shown as solid black, such electrons form a flow of electric current.

AC/DC explained

Electric current flows in two main ways. Direct current (dc) involves electrons flowing in one direction - from negative to positive. Its effect is like a band-saw, in that work is performed by the blade moving constantly in one direction.

With alternating current (ac), electrons reverse direction (in a 50 Hz grid network at 100 times a second). Each complete cycle is thus completed 50 times a second. Action and effect is like a cross-cut saw: i.e. similar work is performed in each direction. In some parts of the world, e.g. Canada and the USA, it is at 60 Hz.

Voltage, current & resistance

Materials that resist electron flow heat up as electrons are pushed through them. This effect can be useful. It is how electric heaters work. But resistance can be undesirable unless heat is required: it causes energy losses. Increasing voltage (the 'pushing force') increases current flow. Larger diameter cables ease electron flow. A longer cable has increased resistance. It needs to be larger, or have a higher voltage, to maintain the same current flow. Ohm's Law explains these relationships.

Energy & power

Energy is a measure of the ability or capacity to perform work. It was originally estimated by James Watt, around 1780, that a brewery horse could lift 33,000 pounds one foot per minute. The amount of energy required to sustain that amount of work was referred to as 1 horsepower (hp).

The concept (of work performed) later became expressed in joules. From that was developed the mks (metre-kilogram-second-ampere) system of which the fundamental quantities are length, mass, time and electric current. (The 'metric' system is based on the metre-kilogram-second, using decimal multiples and submultiples as necessary.)

The International System of Units (SI) is based on length, time, mass, electric current, temperature, luminous intensity and amount of substance. The units are the metre, second, kilogram, ampere, kelvin, candela and mole (respectively). It is virtually the world standard.

The watt later replaced horsepower as the unit of energy (work done). It is equal to one joule per second. The world generally agrees 745.7 watts (usually rounded up to 750 watts) equals one horsepower. It relates to electricity in that one joule per second equals 1.0 watt - as does 1.0 volt times 1.0 amp. America and France, presumably having inferior horses, define 1.0 hp as 735.5 watts.

Power is the rate at which energy is generated/used. It is expressed in watts. The measure of the quantity of energy generated or used over time is measured in watt hours (Wh). It can also be expressed in joules/second but more commonly (electrically) in amps times volts (i.e. watts).

Wattage is thus a measure of the rate at which work is performed and the rate at which energy is consumed in performing that work. The latter always exceeds the former. An '800 watt' microwave oven generates the equivalent of 800 watts in heat but at a typical 55% efficiency, draws 1250 or so watts. Fed via an inverter in an RV, it consumes about 1500 watts.

Power factor

That one watt equals 1.0 volt x 1.0 amp is always so with direct current (dc). With alternating current (ac), however, except with a resistive load (e.g. an electric heater) that acts as if the current were dc. The reason is that loads such as fluorescent lights, electric motors, battery chargers, cause the voltage peak to lead or lag the current peak.

Figure 2.2. Rowers work more efficiently if pulling at exactly the same time. Alternating current acts like that too (with volts and amps). Pic: 'Concentration' © dreamstimes.com.

The effect is like rowers slightly out of phase. Each does the same amount of work as if they were synchronised but the boat will not move as smoothly or as fast. Extra energy is needed to 'fill the gap'. (Figure 2.2.)

In electrical systems, that energy is not consumed. It is 'borrowed' from the supplier and cyclically fills the gaps. This effect, called 'power factor' (the abbreviation is PF), is the difference between the apparent power and the effective power. It is shown as a digit between 0 and 1.0. For many ac loads, up to 30% more current must be available than used e.g. approximately 0.7 PF.

Power factor does not occur with purely resistive loads such as heating. A 2.0 kW generator can thus run a 2.0 kW bar fire but not an induction motor larger than 1500 watts or so. Adverse power factor can limit the output of a 2000 watt battery charger to a probable 1300-1400 watts. Capacitors can be added to battery chargers, motors, etc., to reduce power factor loss - to a typical compromise of 0.8.

As the nature and power factor of the load is often unknown to generator makers, they often quote output in volts x amps. Motor manufacturers specify power in watts, and the energy required to achieve that in volts x amps (VA) or kilovolt amps (kVA). Power factor is a major problem for power suppliers as it requires the generating and supply network to be 25%-30% larger than needed.

Chapter 3

Overview of an RV's electrical needs

There are two main approaches to RV electrical power. The first, mostly in the USA, is to stay only in 'trailer parks' where ample and reliable 120-240 volt (60 Hz) energy is available, and to rely on alternator power to run a fridge while driving. In Australia, however, increasing numbers of people camp wherever they can. Many stay in caravan parks only occasionally. Some never at all.

A few purists still do without electricity except for dry battery powered torches and possibly a radio, but the majority seek at least the basic electrical facilities they have at home. A few would like them all. There are various approaches to the above, and even the latter can be provided - at a cost.

The early days

Figure 1.3. Dynamo (from 1930s Alvis Speed 25). Pic: Car & Classic Ltd.

Early systems relied on energy held within pre_charged batteries used only for lighting. They used (kerosene-powered) fridges that were only marginally adequate. This enabled a few days stay on-site, until the batteries became discharged.

Direct current dynamos (Figure 1.3) of often 6.0 volt for running the car's electric lighting, etc., were available by 1911, but sometimes only as an 'after-sale' option. Until 1918 or so, many car makers resisted attempts to use them at all.

The dynamo assisted recharging and powered the on-road lighting, but its limited output (and hence charging ability) required several hours driving each day to generate even the modest electrical power of early caravans and motorhomes.

*Figure 2.3. This 1920s Angela caravan needed no electrical power!
Pic: Dennis Publishing.*

Magnetos (that provided ignition as the engine turned over) were used until the mid-1920s. Coil ignition was introduced by Delco's founder, Charles Kettering (initially for Cadillac) in 1910. This remained much the same until the major introduction of the alternator (by Chrysler) in 1960. At much the same time, the output was increased - to 12 volts.

The alternator's output was rigidly controlled by an associated regulator that caused the alternator to maintain a constant (typically) 14.2-14.4 volts. Its most vital task was to ensure the starter battery had sufficient charge to provide sufficient current for the starter motor.

Today, starting big 4WD engine requires 500-600 amps but only for two to three seconds. This depletes the starter battery by only 2%-3% of that battery's capacity (typically less than two amp hours). The alternator replaces that in only two to three minutes (Chapter 12).

While the vehicle is stationary, that battery also energises interior lights, electric door locking, immobiliser warning light, and the electric clock but these, collectively, draw little more than the battery loses in self-discharge.

The alternator's major role nowadays is to provide power for all of a vehicle's ever-increasing on-road electrical loads.

An alternator's generally surplus output enabled RVs to charge battery banks capable of running reasonable auxiliary needs. This, however, began to change in the late 1990s, when computer engine management systems were introduced.

Alternator-derived energy was increasingly needed for additional functions as well, e.g. electrical power steering, suspension stability systems, etc.

Using the alternator for charging auxiliary RV batteries became increasingly controversial. It was deemed to adversely affect the operation of those systems. It also raised warranty issues.

From 2001 or so, some alternators were temperature controlled. They ran at too low voltage to adequately charge batteries. This trend continues: most post-2014 vehicles have variable voltage output alternators that often drop below 12.3 volts and only increase when the (starter) battery requires charging to a now 80%.

It is possible that some RVs *may* eventually have alternator output being substantially or totally confined to the original vehicle use, but it is currently (late-2019) too soon for anything but speculation. If/when that happens, fuel cells (Chapter 21) are likely to be an adequate replacement.

While lighting and TVs have become increasingly efficient, microwave ovens, and pre-2014 colour TVs, etc., still gobble power, resulting in many bigger RVs drawing more electrical energy than previously.

For most vehicles, dc-dc charging (Chapter 13) isolates the RV's 'house' electrical systems from the alternator and provides fast, deep and safe auxiliary charging. Chapter 15 covers post-2013 vehicles.

Improved batteries

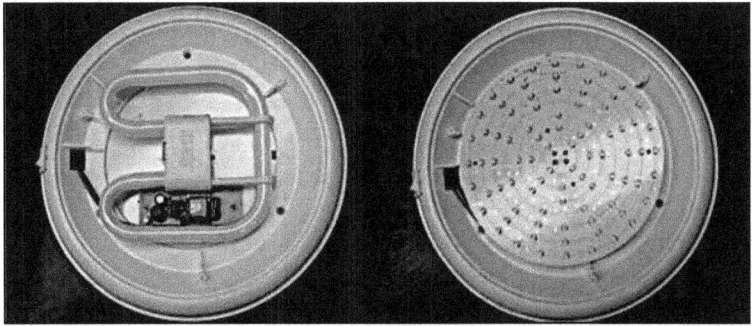

Figure 3.3. Many fluorescent and incandescent light fittings can be retrofitted to LED. The pic on the right shows a fitting converted to LEDs. Pic: ledsunlimited.co.nz.

Around 1980 or so battery technology, that had been stagnant for close to a century, began to improve. Despite this, most practical RV batteries still store only 50% or so more energy (re size and weight) than in the 19th century. Lithium batteries (Chapter 9) can store up to four times as much but in mid 2019 were still three or so times the price of most other batteries.

Solar module prices fell 75% from 2009 to 2011, and continue to fall. Given space for the modules and associated battery capacity, most electrical appliances that people seek to use in a caravan or motorhome can now be feasibly driven from solar.

As this book shows (and explains how to do), an RV electrical system that is designed and scaled appropriately will enable an RV to remain indefinitely away from mains-power and to do so reliably. It is necessary to think ahead, however because vehicle electrical technology is changing fast. It already requires a new approach in several areas.

Lighting

Warm white compact fluorescents are practical and economical. They have internal electronics that run at high frequency, and by so doing, avoid the slight flickering of full length fluorescent tubes.

Ultra-efficient LED globes are replacing other forms of RV lighting: they are the only ones to consider for new installations. They can usually be retro-fitted (Figure 3.3).

Appliances

Until recently it was often more efficient to use 12 volt dc appliances rather than an inverter's 230 volts (an inverter converts a low dc voltage to grid voltage ac). Inverter efficiency and performance, however, is now a typical 93%-96%. A few are even higher.

Using 230 volts ac all but eliminates voltage drop in RVs, and enables a wide choice of good and affordable products. These vary in efficiency but the 'Energy Star' ratings system enables easy assessment of the main previous heavy gobblers - particularly fridges, and items (such as cooling fans) driven by induction motors.

Away from mains-supplied power or power from a generator, electrical usage is best limited to lights, radios, DVD players, laptop computers and communications equipment, iPads, TVs, fans, blenders, water pumps, fridges, etc.

That which cannot realistically be run from small scale solar and/or alternator power, is anything that, as its intended function, generates heat over time. As there's more energy stored in 9.0 kg of LP gas than can be held in 1000 kg of lead-acid batteries or 250 kg of lithium batteries, LP gas (or diesel) is still best used for ovens, grills and water heating.

Microwave ovens are fine in large RVs but borderline in smaller ones with limited available energy. An '800 watt' microwave oven produces 800 watts in 'effective heat' but draws about 1200 watts, or about 1330 watts if run via an inverter. It needs an AGM battery of at least 150 Ah. Ten minutes usage may be half a day's energy draw in a small RV. It is possible to accommodate this but it makes more sense to use the microwave oven *only* where 230 volts mains-power is available. If not, that under $299 unit may cost $1000 more in solar and battery capacity.

CPAP

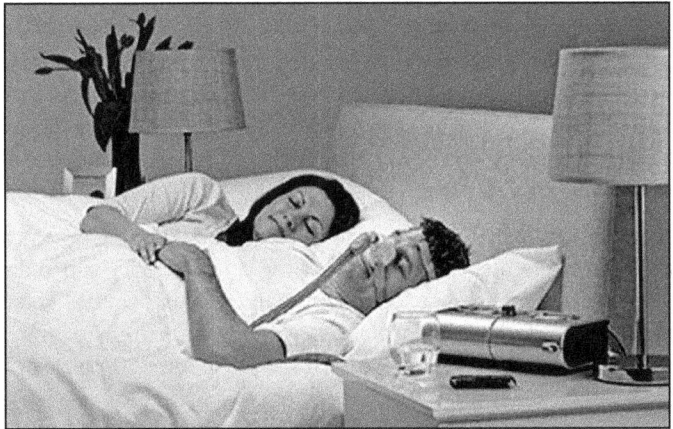

Figure 4.3. A CPAP in use. Pic: ResMed (USA).

These continuous positive airway pressure devices are used by people who suffer from snoring, sleep apnoea and other forms of sleep-disordered breathing. They are largish volume, low pressure air pumps, some of which incorporate humidifiers. Some also heat the supplied air.

Until 2002 or so these units drew from 130-300 watts. This was really too high for realistic 12/24 volt power excepting in large coaches with equally large battery banks. Their efficiency since then has improved greatly.

By and large CPAP energy draw is related to the treatment pressure: typically from 590 Pa (6 cm H^2O) to 1960 Pa (20 cm H^2O). The more efficient (non-heated) units draw (at 12 volts) 0.65 to 1.4 amps for the above, but a few older ones draw twice that. Those with heated humidifiers draw several times more.

Some have inbuilt inverters but most are 230 volt ac units that require a high quality pure sine wave inverter if run from 12/24 volts.

CPAP machines may have a poor power factor (Chapter 1 and 2). If so, these require an inverter that has an output of 50% or so higher wattage than that of the machine.

This is a very specialised area. Anyone with (or suspected) sleep apnoea syndrome, sleep disorder, snoring or similar condition should consult a doctor or other specialised professional in this area. Advice on specific CPAP machines should be obtained directly from the major vendors in this field.

Refrigeration

An electric refrigerator typically accounts for 50 to 70% of an RV's electrical consumption. The larger RV fridges need 250 to 350 watts of solar capacity to drive them and, at minimum, a 200 Ah battery to run them at night (particularly in tropical areas), plus a fuel cell or generator for times of little sun. Solar is now cheap, and thus practicable for RV use if there is space for the solar module area required.

Three-way (12 volt/mains-voltage/gas) refrigerators running on gas are less convenient and while more costly, the saving on electricity approximates their higher initial cost. They use about 0.45 kg of gas per day. Their 12.5 to 25 amp draw from 12 volts precludes running on battery power except while driving and for short roadside stops. They need 230 volts, or to be run on gas whilst on remote sites.

Some users are prejudiced against three-way fridges as early models were ineffective in extreme heat. The later ST and T-rated units work fine in Australia if installed correctly.

Water pumps

The 230 volts ac water pumps draw too much energy for camper van and motorhome use. Even those supplying only one or two taps draw about 500 watts, and twice that while starting. RV 12/24 volt pumps typically draw about 60 watts (about 5 amps and 2.5 amps respectively). Some made recently are quieter and provide smooth constant flow - yet cost no more to buy. refers - and also shows manufacturers' typically claimed pump draw.

Air conditioning

Excepting in large RVs with ample space for solar modules and substantial battery capacity, it is not as yet feasible to run an air conditioner for more than a few hours during the night from battery-stored solar. This situation may well change because the very best small (domestic) reverse-cycle air conditioners are becoming increasingly more efficient.

As of 2019 the Daiken US7 (2.5 kW unit rated reverse cycle unit) draws only 420 watts on its cooling cycle. If run via an efficient inverter, this corresponds to about 33 amps at 12 volts. This is still marginal for solar but a few costly USA units are now claimed to use even less.

Converters

Most production RVs are now fitted with a so-called converter that, when 230 volts is connected to the vehicle, provides an unregulated 13.65 volts directly to power the RVs 12 volt needs.

These units present major problems for those intending to stay away from mains power for more than one overnight stay. This issue is covered on Chapter 11.

Imported RVs

There are known to be major electrical compliance issues with some US imports. A loophole in the regulations enables the original buyer (only) to use them in Australia by adding a 230 to 110 volt transformer. Many owners assume (wrongly) that this confers electrical compliance. It does not. To be 100% compliant they may not have any form of 110 volt anything - from wiring to all appliances.

Before being offered for sale, such units must be brought totally into full electrical (and gas) compliance. Take this seriously. Selling a non-compliant RV is a criminal offence in some states of Australia and a civil offense in others. This issue is covered in depth in Chapter 40.

Chapter 4
Providing the power

There are various approaches to implementing an RV electrical system. These range from total dependence on substantial 230 volt power, to settling for minimal but comfortable needs rather than all that one would like to have. Most settle for somewhere in-between but commonly find that the more they travel, the less they need.

No single approach needs be followed exactly but unless you really do know what you are doing, it is best not to depart too much from that described in this book.

Always check that what you are planning is physically possible. Is there space and weight carrying capacity for the batteries? Is there enough space to mount solar modules? Does an old energy gobbling fridge need replacing? Current top quality fridges us far less power so the cost of updating will be recovered by the need for less battery capacity.

Increasing wiring size, particularly on older RVs, always assists and may be possible without removing and replacing exterior cladding. If not, new wiring can often be concealed within cupboards and under seating, etc.

Caravan park power

Caravan parks are legally required to have one 15 amp outlet per powered site. Except for rare full size electric ovens, that 15 amp outlet will supply all of a typical RV's electrical needs.

A few really big rigs have twin 15 amp systems, often necessitating their owners to rent a second site to obtain that extra supply.

Staying routinely in caravan parks, however, is lessening as RVs become increasingly self-contained, yet owners may still be charged $50 or more for an overnight stay - even if they need only a plot of grass - and no 230 volt power.

Away from mains power

What grandpa did, and some of all ages still do, was to leave home with a hopefully-charged auxiliary battery and drive to the campsite. He'd use that until it was deeply discharged and attempt to recharge it or use candles or a Tilley paraffin (kerosene) lantern. Or drive home.

Figure 1.4. This lovely old original Teardrop relied on gas and paraffin.
Pic: Hilary Walker

Dynamos back then (prior to alternators) had very limited spare capacity but as the load was mostly lighting, it was feasible to use them to charge a small auxiliary battery.

That battery was normally isolated from the starter battery by a manually operated switch. Switching it off tended to be overlooked but was not a major problem if sufficient voltage remained for the spark coil ignition because all cars back then could also be manually started.

Some had magneto ignition (a high voltage dynamo) that needed no external electrical power, so even flat batteries were of no concern when starting an engine manually.

The 1960s introduction of alternators enabled larger batteries to be charged - but not that effectively because fixed voltage alternators precluded charging much beyond 75%.

This changed in the mid-1990s, when the decades-old multi-stage charging technique began to be used for RVs, and even more so by about 2009, by the adoption of dc-dc alternator charging (described inChapter 13). The latter enables a 40% to 50% discharged 110 amp hour conventional deep-cycle battery to be close to fully-charged inside three or four hours driving. If alternator capacity allows, AGM and gel cell batteries of that capacity will likewise charge in two and a half to three hours driving.

For quick 'around Australia's, the above is a cheap and simple approach. Whilst their life is unlikely to exceed 12-18 months the batteries made to power golf carts, fork lift trucks, etc., will usually suffice.

Staying longer on-site

For optimum battery life, all but lithium auxiliary batteries needs to be fully charged on most days. This requires that average energy draw (including inevitable losses) must not exceed the average energy input by even a small amount, or a deficit builds up.

If the deficit is due to inadequate alternator or solar generating capacity, adding more battery capacity (as many people do) is worse than useless. If electrical input is not being generated it is not there to store. Adding battery capacity simply increases total battery internal losses.

In such situations reduce the load, drive longer or add more solar. Never increase battery capacity unless the existing batteries reach full-charge by 1 pm to 2 pm most days wherever you are.

Recommended approach to battery and solar required

The following, and by far the preferred and totally proven way of achieving that outlined above, works reliably in conditions that prevail across much of Australia, except down south in mid-winter. Even there, it is fine if you have a correctly installed three-way fridge running on gas while on-site.

Figure 2.4. Self-sufficiency as defined by Mr. McCawber (Charles Dickens): "Annual income twenty pounds, annual expenditure nineteen nineteen six, happiness happens. Annual income twenty pounds, annual expenditure twenty pounds ought and six, result misery." Solar is like that too!

The essential requirement is that, on most days in the area where you intend to travel, you have sufficient solar capacity to be able to charge the battery close to 100% by 1 pm to 2 pm. Chapter 11 shows how to assess this. If you follow this, apart from during rare freak days, the battery can be relied upon to supply the deficit over a few days of little sun. This is because irradiation only rarely drops below 35% to 40% of normal during most of the year. This frees you from ongoing concerns about power. Further, lithium and most other batteries should last four to five years. Some owners obtain more.

If you can accommodate the solar capacity required, it is feasible to do the above with bigger fridges and fridge/freezers. If not, the deficit can be compensated for by having an inverter/generator driving a three-stage mains battery charger run from the generator's 230 volt ac outlet.

For deep-cycle lead-acid batteries the charging rate is ideally 15% of overall battery Ah capacity, or 25% to 35% if the battery is a gel cell or AGM. LiFePO4 batteries will accept far higher charge rates.

If you use a three-way fridge, an almost bullet-proof system is feasible using two or three 120 or 130 watt solar modules and 100 to 200 amp hour battery capacity.

From information in the following pages, you can readily tailor a system to suit your own needs and finances. It can be a totally self-sufficient system or one that enables it to be charged by the vehicle's alternator or topped up by a quiet inverter generator, or a (currently costly) fuel cell, during a longer term lack of sun.

The above approach can still be used with RVs that have electric fridges of around 220 or so litres, and also microwave ovens - but requires a lot more solar energy.

Battery capacity

Unless you have a generator, limit battery capacity to that brought close to 100% charge by noon (on site) most days or add more solar until you can.

Fifth-wheel caravans (i.e. caravans that have the tow hitch above the tow vehicle's rear axle) can usually cope with the weight of more batteries, but there is no point in having that battery capacity unless you have enough solar to charge it.

If you intend to spend time in southern areas during June to August, it makes economic sense to recharge the battery bank via a small quiet generator driving a 230 volt charger. Only consider alternator charging alone if you routinely drive for sufficient time to fully charge the batteries.

By using enough solar, total electrical self-sufficiency can be achieved silently and pollution free. It has to be competently designed and installed but, if it is, will prove more reliable than mains power. If not, it may cause ongoing problems until fixed. Doing it properly costs more initially but, as batteries then last much longer, less frequent replacement goes part of the way toward the one-off cost of adequate cabling and solar capacity.

A large number of RV and cabin solar systems are close to self-sufficiency but if, on average, there is an ongoing energy deficiency, however small, that system will eventually run out of power.

Never cut back on solar capacity. If/when more affordable fuel cells (Chapter 21) are available, consider using one for back-up, not least as national parks and many caravan parks ban even quiet generators.

Solar capacity

The following sections show how to scale solar capacity. From thereon it is simply a matter of selecting whatever combination of solar modules is required.

Mounting is eased by using identical sized modules, but it is electrically fine to connect solar modules of even hugely different current output in parallel (plus to plus, minus to minus) to increase overall capacity. They must, however, be of similar voltage.

It is also fine to connect solar modules in series (consecutively plus to minus) to increase voltage as long as the solar regulator permits (Chapters 22 & 24). Here, the modules must all be of identical current output as current output is limited to that module which generates the least. Voltage is directly additive.

Always err on the side of too much solar capacity - never too little. Solar capacity is now so cheap that the limit for many RVs is roof space and/or weight.

Other sources of power

A 1.0 kW inverter/generator can produce a constant 800 watts. An appropriately-rated multi-stage charger operating from the 230 volt ac output of that generator is likely to input up to 35 amps - providing your battery bank can accept that rate of charge.

Most AGM batteries can do so, as will conventional lead-acid battery banks over 150 Ah.

Lithium batteries are claimed to charge safely at three or more time their amp hour capacity: e.g. a 100 amp hour such battery can be charged at 300 amps.

There are various types of lithium batteries, but as explained in Chapter 9, those used in cabins and RVs use a very safe lithium-iron technology chemically abbreviated as LiFePO4.

Chapter 5

Scaling the power required

The examples in Chapter 38 show various systems. Check these before making firm plans, particularly the last but one example, that shows how not to go about it! That done, start initial planning by listing realistically usable appliances you feel you must have.

Table 1 - typical energy draw

Item	Watts
Air-conditioner	1000-2500
Blanket (under)	60-120
Blender	150-350
Bread maker	350-450
Can opener	100
Cassette player	30
DVD player	30
Clothes dryer	2400
Coffee grinder	75
Coffee percolator	500-600
Compact fluro	8-18
Computer (desktop)	200-350
Computer (laptop)	20-30
Computer printer (inkjet)	70
Computer printer (laser)*	1250
Dishwasher	1000-3000
Fans	25-100
Food mixer	450-550
Fry pan	1200
Hair dryer	800-2000
Iron	1200
Juicer	350
Kettle/jug	850-2000
Lights (LED)	3-8
Macerator	300
Microwave oven (800 watts)**	1500
Mobile phone (charging)	10-20
Radio	15-50
Sewing machine	75-100
Stereo	50-60
Toaster	500-1500
TV (32 cm LED/LCD)	40

Item	Watts
Vacuum cleaner	700-1400
DVD	30
Washing machine	200-600
Waste disposal unit	500-750
Water pump (12/24 volt)	50
Water pump (230 volt)	>750

*If used off-grid a laser printer **must** be driven by a true sine-wave inverter.*

**Microwave oven ratings reflect the 'cooking heat' they produce - not the energy draw while doing so.*

Chest opening fridges ***	watt hours/day
Autofridge (40/70 litre)	290-480
Engels, etc., (40 litre)	350
Engels, etc., (60 litre)	550
Engels, etc., (80 litre)	750

Door opening fridges ***	watt hours/day
110 litre	900
150 litre	1200
220 litre	1750
300 litre	2400

The above fridge draws apply to 12/230 volt fridges marketed for RV use. A few, but costly 230 domestic fridges have less than half the daily draw of the larger ones shown above. Some are used successfully in RVs. Ditto reverse cycle air conditioners.

****Due to the industry method of rating product, fridge draws are shown here as that typically drawn over 24 hours.*

Energy draw of proposed lights and appliances

Having prepared list of appliances, using your version of Table 2 (Proposed Energy Draw), enter the maker's specified current draw of each item and total the data entered.

If you know only the current (amps), multiplying that by the system voltage (i.e. 12, 24 or 240 volt) is the draw in watts.

Multiplying the draw of each item (watts) by the hours used daily is the energy draw in watt hours.

Estimating a fridge's draw is tricky. Some makers show total daily draw (at a typical 25°C ambient temperature). Others show only the draw when cycled on - Chapter 31.

If in doubt, assume between 0.7 and 1.0 watt hours per day (i.e. volts x amps per day) times the fridge's capacity in litres. If intending to stay for long in hot places allow a 5% increase for every 1.0 degree C above 25°C. (See also Chapter 31 regarding all of this).

Microwave oven ratings reflect the 'cooking heat' they produce - not the energy draw while doing so. Many people overlook this: an '800 watt 'microwave oven will draw 1200-1300 watts.

Table 2 - proposed energy draw

Column A Device/s	Column B Watts	Column C Hours/day	Column D Watt hours/day
Lights LEDs (4)	5	4	20
Laptop computer	30	2	60
TV (32 cm LED)	40	3	120
Water pump	50	0.25	12.5
Fridge (12 volt chest 40 litre)	30	12	350
iPad (approx)	4	2	8
Totalled Column D			570
Add 25% for losses - see Note			142
Final Total			712

Note: re 'Add 25% for losses'. *Most batteries charge at a higher voltage than that at which they discharge. Further, the voltage of conventional batteries drops considerably under load. Assume losses of 25% if using conventional deep-cycle lead-acid batteries, and 15% if AGM and gel cell batteries. For lithium (LiFePO4) battery efficiency use a (conservative) overall correction of 10%. In practice most will be 5%–10%.*

1. In column A, list lights and appliances - of your choice.

2. In B, enter the wattage or the total wattage if more than one device (such as lights) will be used at the same time. Add 10% for any 230 volt devices driven by an inverter or inverter/charger made since 2000. Add 15% if made prior to 2000 or, replace by later more efficient equivalents particularly lighting, fridges and inverters. Retaining inefficient ones will cost you more in extra solar, etc., than that of replacements.

3. In C, enter the hours each is used daily. Use likelihoods, not rare maximums.

4. Multiply each entry in B by that in C - and enter in D.

5. Total and enter all the D entries in the third but last D row - as shown below. As some batteries and their charging processes lose up to 25% in terms of energy, add 25% to compensate (see also the Note below).

Following pages show how to obtain the necessary energy needed to run the system reliably.

Chapter 6

Installing safety and legality

Terms used to define voltage do not mean what many non-technical people believe they mean. They vary considerably from country to country - particularly in the USA. The International Electrotechnical Commission (IEC), however, has rigid definitions of Extra-low, Low, and High voltage that are applicable in many countries including Australia and New Zealand.

Extra-low voltage has three categories but that applicable to most RV systems, is defined as any voltage not exceeding 50 volts ac or (ripple-free) voltage not exceeding 120 volts dc.

Twelve volts may provide a slight tingle if one's fingers are wet (but can cause muscular paralysis if fully immersed), 24 volts does more of the same. You'll be well aware of 48 volts. The 80 plus volts from a 48 volt solar array is dangerous. That of a 72 volts array approaches a potentially lethal 120 volts.

Another common risk at these voltages is of short circuits (caused by live conductors touching) resulting in cables overheating and burning. A slight break in a connector or cable may cause arcing, with subsequent risk of fire. Where vehicles use the chassis and other metal parts as a common negative lead, any live lead that touches such metal causes a short circuit. See also Chapters 7 & 8.

Low voltage is defined in most countries as 50 to 1000 V ac, and 120 to 1500 V dc. All introduce a risk of possibly lethal electric shock. The so-called 'mains voltage', previously defined in Australia/New Zealand as 240 volts, became defined (in 2000) as 230 volts +/- 6% to bring it into line with IEC terminology. In practice it is still 235-240 volts or so.

High voltage is defined as that exceeding Low voltage. Its ability to form and sustain electrical arcs through air adds a substantially higher risk, but mainly only for electricians working in this area.

The USA has many terms for 'mains-voltage'. These include grid-power, household power, household electricity, house current, power lines, domestic power, wall power, line power, ac power, city power and street power. By and large it is 120 volt single phase and 240 volts two-phase 60 Hz ac. The US National Electrical Code, however, defines low distribution system voltage as 0 to 49 volts.

Australia/NZ regulations

Confusion can arise in Australia and New Zealand because many non-electricians refer to that defined as Extra-low voltage as 'low' voltage. No qualifications are needed to install, modify or repair most Extra-low voltage wiring and components but brakes and lighting, etc. are governed by ADRs (Australian Design Rules) and the work must be done by an electrically certified mechanic. Apart from the above, and a few other specialised areas such as medical, no licence is needed to work on Extra-low voltage systems.

With the exceptions below, all Low voltage work (on 'electrical installations') must be done by licensed electricians, who are also responsible for inspection and certification. Energy Safe Victoria, however, deems that RVs are not electrical installations. It advised (in writing) that: 'Because the RV is not an installation a licence is not required to perform the electrical work and a Certificate of electrical safety is not required'. Such RVs must nevertheless meet all requirements of the relevant AS/NZS standards: AS/NZS 3000 2018 and AS/NZS 3001:2008 as Amended in 2012.

New Zealand allows minor such work to be done by 'an experienced person'. It must then be inspected and certified by a licensed electrical contractor. There are also defined areas (e.g. gas fitters installing gas heaters that have 230 volt electrics) where they have limited permission, as long as their work is then inspected and approved by a licensed electrician.

The above still applies for systems run from an inverter or generator - even if there is no provision for external mains connection. Certification is not required for installing inverters with inbuilt outlet sockets enabling 230 volt ac appliances to be plugged into them. (See Chapter 16.)

Chapter 7
Installing 12/24 volt wiring

Cables resist electrons (electricity) flowing through them. They lose energy in the form of heat. Because of this, the voltage at the end of any cable will, under load, always be lower than at its source. Voltage losses increase in proportion to cable length and to the amount of current flowing. They decrease in proportion to an increase in cable cross-sectional area.

Local, UK and EU appliance and battery makers specify cable following the International Standards Organisation's (ISO) convention of rating cable by its conductor cross-sectional area (in mm^2). Mains-voltage cable is rated this way. That, and ISO-rated 12 volt cable can be bought from electrical wholesalers, many marine electrical suppliers and a few specialised auto-electrical suppliers.

In many countries, including Australia and New Zealand, however, the 'auto-cable' sold by auto parts suppliers and hardware chains is, for reasons that defy sanity, rated by its overall diameter including its insulation. It literally indicates only the size hole it can just be pushed through - not the all important size of the copper content!

As auto-cable insulation thickness varies from brand to brand, apart from the 8 mm size, no comparison with any standard or rating method is feasible. As a rough guide, auto-cable less than 4 mm has a conductor area of 0.5 mm^2 to 1.5 mm^2. The 4 mm size will have a conductor area between 1.8 mm^2 and 2.0 mm^2, 6.0 mm is typically 4.7 mm^2. The 8 mm size auto-cable is close to the ISO equivalent: typically 7 mm^2 to 8.0 mm^2.

DIY installers are likely to use auto cable. It is a good product electrically and, as long as the rating is understood, there is no reason not to use it. However, not knowing about auto-cable's misleading rating results in massive voltage drops that may plague an RV for life. Auto electricians say most RV electrical problems (especially with battery charging and fridges) have this common main cause.

Auto cable is also a purchasing trap. If you go to almost any auto parts store and even ask specifically for (say) 4 square millimetre cable it is all but certain you will be sold 4 mm auto-cable. It may *look* the same but is mostly plastic. Cheap thick jumper leads are an extreme example.

Current ratings

Compounding auto-cable 'rating' madness is that it is also sold as '10' amp, '25' amp', etc. This relates only to the safe current-carrying capacity of the cable and that varies with the type and melting point of its insulation. Current ratings do not relate to voltage drop. A '25' amp cable may well have an acceptable voltage drop over two metres but across 10 metres that drop may be 1.0 volt.

For the cable sizes recommended, no allowance need be made for heat build up for 12/24 volt wiring. This is not the situation for 230 volt circuits if covered by heat insulation (see Chapter 8).

High quality inverters are designed to handle up to twice or more their rated output for some seconds. A 1500 watt inverter may thus briefly pull over 250 amps. This necessitates short and heavy connecting cables and being located as close as possible to the battery bank. Select cable that ensures no more than 0.2 volt drop using the method shown in Table 3 in this Chapter.

For any given wattage, current is halved in a 24 volt system. As it is percentage volts drop that matters, a drop of 0.4 volt is acceptable. Thus 24 volt cables need be only 25% that of 12 volt sizes.

AWG/B&S

The USA (mainly) uses AWG (American Wire Gauge), or the all but identical B&S (Brown & Sharpe) logarithmic-stepped standardised wire gauge systems. There are 44 such sizes ranging from 0000,000, 00, 0; and then 1 to 40. Those from 0000 to 0 are often shown as 4/0 to 3/0, etc. Most vendors stock only the 'even' sizes but 'conversion' charts show the odd numbers as the next smaller even-number cable. Such rounding-down adds a typical 25% higher voltage drop. This is serious if it occurs in an already inadequately cabled circuit. The comparison chart in this chapter (Table 4) shows that only a few ISO sizes directly correspond to AWG/B&S 'even number' sizes. If you *can* obtain the correct odd-number size, then do so. If not, use the next larger AWG/B&S. For example, for ISO 2.5 mm² use 12 AWG/B&S unless you can obtain 13 AWG/B&S.

Cable size needed

Many auto electricians accept a loss of 0.5 volt for a vehicle's legally required lights, etc., but that is hugely too high for charging auxiliary batteries and for RV wiring generally. The preferred 0.15 to 0.20 volt drop (at 12 volts) and 0.30-0.40 volt (at 24 volts) is readily achieved across the short cable runs in camper trailers. It is feasible, but needs heavy cabling in long caravans and motorhomes.

Table 3: voltage drop formula

$$\text{Drop in volts} = \frac{(L \times I \times 0.017)}{A}$$

where:

L = total conductor length (m)

I = current flowing (amps)

A = cross section of cable (mm²)

ISO Standards use a constant of 0.0164, but 0.017 is easier to remember. The difference is too small to matter.

Ignore voltage drop/cable size charts. Unless indicated to the contrary, they are for voltage drops of 0.5 volt. This is *far* too high for 12 volt RV wiring, excepting for LEDs. These charts also introduce gross errors as they show the same cable size for lengths (say) between one to three metres, and wide spreads of current.

The simple voltage drop formula shown here indicates cable size for any length, current, and voltage drop. Use only this. Ensure the 0.15 to 0.20 V drop is across the entire conductor run - from battery terminals to the device concerned and back again. For example, if a fridge is three metres cable length from the battery, that is six metres of conductor (i.e. not three metres).

A simple approach is to buy cable in only three or four main cable sizes, accepting that some runs may have cable heavier than needed. This saves money: cable is far cheaper when bought in rolls.

Copper cable eventually tarnishes in most environments. It is better to buy the more costly tinned copper cable (it is actually copper electroplated with nickel alloy). Such cable is obtainable in ISO sizes from most boating suppliers and also from Springer Low Voltage Electrics (Qld).

For inverter and other really heavy current cable, second-hand welding cable can be often be bought for bargain prices. It has a vast number of fine strands and is perfect for battery, starter motor and winch cables. Welding cable, however, has a tendency to attract moisture drawn in by capillary action so cable ends need to be well sealed with a silicon preparation and heat shrink tubing.

Earth return

Two separate conductors are required to conduct direct current. One is positive (+ve), the other negative (-ve). Using the vehicle's chassis as the negative conductor provides a low resistance path but only works well if the earth connections are clean and tight.

On vehicles with variable voltage alternators (Chapter 15) all RV negative leads must go the chassis - not to the battery earth terminal. This is because the main cable from the chassis to battery negative and/or the alternator negative acts as a so-called 'current shunt' and measures the current flowing.

Table 4: Cable sizes compared

AWG/B&S	ISO	Autocable *
18	0.75	3
17	1.0	
16		
15	1.5	
14		4
13	2.5	
12		5
11	4.0	
10		6
9	6.0	
8		8
7	1.0	
6		
5	16	
4		
3	25	
2		
1		
0	50	
2/0	70	

Table 4: true comparison between ISO and AWG/B&S. The ISO sizes refer to cable cross-sectional area in mm². Most tables 'round-up' the numbers - such that 1.0 mm² becomes 18 AWG instead of its actual 17, introducing errors of up to 25% and can result in voltage sensitive cables being under-specified. (* Approximate)

As with most cable charts, the one(s) in this book shows cable rating data for single conductors. Specify anything (say) three metres from the battery as if it were six metres of single cable. This will be overkill if earth returns are well done but experience shows that, in RVs, few are.

Tracking voltage drop

Tracking voltage drop requires measuring small voltages across cable runs while they are carrying their probable loads. This necessitates using a digital voltmeter or (preferably) multimeter.

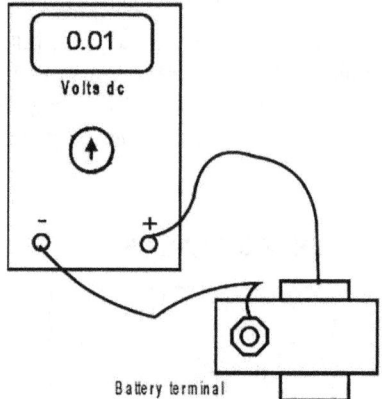

Figure 1.7. Checking for voltage drop between a battery post and connecting lug. Reading, on full-charge, or heavy discharge, should be close to zero. Pic: rvbooks.com.au

Check the battery terminals first. To do this, turn on all lights and appliances (or the microwave oven if you have one) that are likely to be on at the same time. Leave the fridge door open - and the sink tap running (to keep both running).

Clean the top of each lead battery post, then press one meter probe firmly into the post and the other into a clean part of the outer side of its connecting lug. It should show virtually zero. A reading above 0.05 volt will be due to a loose terminal, poor crimping and/or corrosion. Correct it and re-check (Figure 1.7).

Next, check each circuit from that battery post to as close as you can get to the appliance whilst (if applicable) it is working at its maximum load. If a drop of over 0.2 volt shows up (i.e. 0.1 volt on either of the two conductors, or 0.2 volt total), clean the contact surfaces of all associated connectors and check again. If there's still a drop, check directly across that connector. It is common to find the cause is inadequate cable size, but it can also be due to inadequate crimping, poor earth contact, a corroded fuse holder or a poor switch contact.

Voltage drop can also be checked by measuring directly across the battery, and then directly across the appliance, but does not pinpoint where problem/s lie. Cables, connections, fuse holders, etc., that are slightly warm to the touch are also an indicator of voltage drop. This measuring method applies equally to charging circuits but as this can only be done with the engine running, care must be taken to ensure that a meter lead does not get caught in the fan belt. Unless truly experienced, enrol the aid of someone who is. Or have an auto electrician do this for you.

Voltage drops of 0.5 volt plus are not uncommon, but excepting for inverter dc feed cables, where over 0.2 volt is inevitable on full load, end-to-end drop across the totality of any circuit should ideally not exceed 0.15 to 0.20 volts at 12 volts, and 0.30 to 0.40 volts at 24 volts, when that circuit is fully loaded. For battery and associated charging circuits and fridges, low voltage drop is essential.

Cable protection

It best not to run internal 12 volt cabling inside conduit. It looks neat and professional and protects against damage, but it is hard to make subsequent changes. Plastic ties, or spiral wrapping are more practicable.

See also Chapters 7 & 8 regarding physically separating 12/24 volts wiring (and conduit for 230 volt wiring).

Circuit breakers

Circuit breakers protect cables if anything causes the cables to conduct excess current, overheat and possibly set on fire. This requires the circuit breakers to be installed close to the battery (but not within its enclosure). Circuit breakers are available from 5 amp upward. They cut off the current flow just above their rating.

One high current main circuit breaker may be used to protect all of an RV's cables, with lower current rated circuit breakers for dedicated groups. For example having individual circuit breakers to protect left and right side lighting circuits, another circuit breaker for the water pump and another for external circuits, etc. Circuit breakers can also double as switches. Locate them as close as possible to the power source.

Good circuit breakers (Figure 2.7) cost $50 plus but may save your RV from burning. The best have magnetic or hydraulic operation. They have ratings from 5 to 500 or so amps.

One manufacturer (Clipsal) states that its ac circuit breakers can be used for dc (at up to 48 volts) but will then have different tripping performance. As dc (but not ac) forms an arc when broken it is advisable to buy only dc circuit breakers as these have much larger contacts that better withstand that arcing.

Figure 2.7 Circuit breaker. Pic: Carling

Cheap circuit breakers are temperature sensitive, they typically trip at 10% less current for (roughly) every 5°C above 27°C. Summer heat in (say) central Australia may cause them to trip at well below their rating and even lower if they are also installed close to a hot engine. As with cable, this is not an area in which to attempt to save money.

Fuses

Fuses protect appliances and the electrical supply from overloading (by melting and thus breaking current flow) in the event of a fault in the protected appliance causing it to draw excess current. They are therefore installed close to (or within) such appliances - Figure 3.7.

General-purpose fuses blow at about 130% of their rating but, as with thermally actuated circuit breakers, are temperature dependent. Fuses should generally be selected such that they blow at currents about 30% to 50% higher than the draw of the appliances they protect.

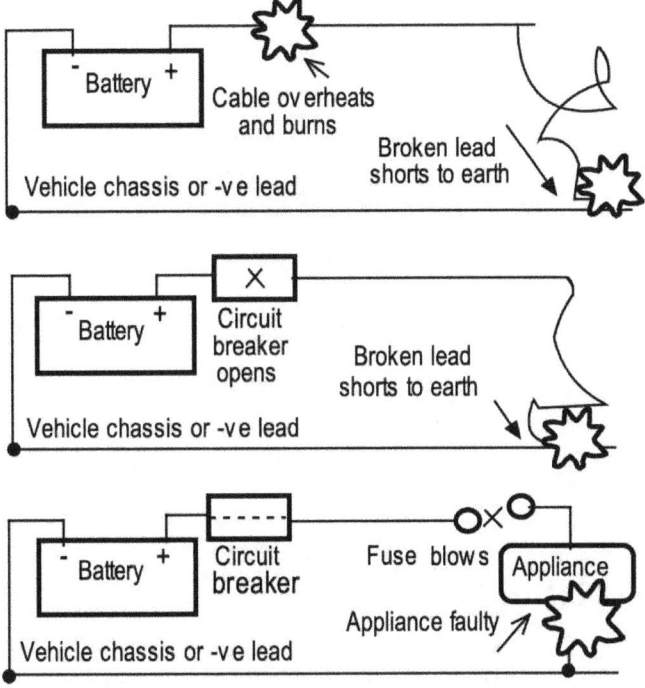

Figure 3.7. Top: Broken/loose cable touches metal and shorts to earth - cable overheats and sets on fire.
Centre: Circuit breaker senses excess current in shorting cable and shuts off power.
Lower: Fuse blows, cutting power to protect appliance against further damage.
All pix: rvbooks.com.au

'Slow blow' fuses are needed for units that include an electric motor. Most electric motors draw twice or more their running current when starting. Some appliances have internal fuses, requiring dismantling for access. If you know how to, consider shorting out the original and replace it by an external one of similar rating.

The previously common tubular glass fuses often lose full contact with their holders. This causes heat that melts the fuses or causes them to blow prematurely. Blade fuses are generally more reliable. They are made in two main physical sizes. For 20 amps and above use only the larger size.

If funds permit, use circuit beakers rather than fuses. Since 1900, insurance records show that circuit breakers provide safer fire protection. Most circuit breakers also double as main on/off switches.

A 1500 watt, 12 volt inverter draws 120 to 130 amps while feeding an '800 watt' microwave oven. The required circuit breakers are readily available from electrical suppliers.

Figure 4.7. Cartridge fuses bolt onto associated insulated bases. Pic: Fuseco (USA).

An acceptable compromise is to protect high current cables by cartridge type (bolted in) fuses. Use the fully enclosed type shown in Figure 4.7, the open ones go off like fire crackers when they blow. Replacement cartridge fuses are not cheap. You may, over time, spend as much on replacing them as the cost of a good circuit breaker.

Winch solenoids

It is worth considering installing a heavy duty solenoid to parallel connect house and starter batteries to power a winch. (A solenoid is a switch used to connect and disconnect very high current using an far lower current to do so - an example is the engine's ignition switch). For winches, that solenoid needs to switch about 500 amps.

Voltage-sensing relays

To ensure reliable starting, these relays delay connecting the RV's auxiliary battery across the starter battery until the alternator is charging at about 13.6 volts. This is typically 2-3 minutes after the engine starts (it depletes the starter battery by only 2% or so). The relay also disconnects the auxiliary battery if the starter battery drops below 12.6-12.7 volts. They are mostly used on pre-2014 vehicles.

Variable voltage alternators

These became used in many new vehicles post 2013, as part of EU emission regulations. This topic is covered in Chapter 15.

Plugs & sockets

Most cigarette lighter plugs and sockets lack mechanical locking. They are fine if used for their original purpose, powering GPS devices or charging mobile phones, but not as permanent power outlets. Electrical contact may degrade over time, resulting in their becoming hot enough to ignite nearby material when current is drawn. Good locking plugs and sockets are made by Hella, Engel and Blue Seas, but it is better to hard wire appliances in permanent use, or otherwise use an Anderson plug and socket (Figure 6.7).

Figure 5.7. Hella sprung locking plug handles 16 amps at 12 volts. Pic: Hella.

Cigarette lighter outlets in vehicles are rarely connected by cable intended for any but short time usage. Their cable's voltage drop may be unacceptably high if used to drive (say) a large chest fridge.

Figure 6.7. Genuine Anderson connectors are excellent products - but be wary of the many ultra-cheap copies. Pic: Anderson.

Especially to be avoided are the miniature two-pin, push-in, chrome-plated plugs and sockets stocked by some boating suppliers. They look good but lack mechanical locking. They are bad enough when new, let alone after a few years. Excellent but bulky, two-pin extra-low voltage plugs and sockets are made by Clipsal. Also excellent are the more compact units made by Bulgin for marine use.

Extra-low voltage switches

Switches for dc operation have heavier contacts and a faster break action that reduces contact arcing. They are made by Hella and a few other companies. In practice, if they are limited to 20% of their ac rating, 230 volt ac switches (such as light switches) work reliably at 12/24 volts. Most work reliably at up to 2.0 amps dc - so are perfect for switching LEDs.

Switch & meter panels

Rather than attempting to make cut-outs in fixed panels, mount circuit breakers, switches, meters, etc., on a separate removable panel. This simplifies initial construction and wiring, and eases subsequent changes as one can readily make up a new panel. If possible, locate the panel for minimum length cable runs, particularly for cables carrying heavy currents.

Current shunts

A current shunt (Figure 7.7) provides an indirect measure of flow current. It consist of one or more short metal rods or strips connected in series in a main battery lead. The shunt introduces a slight resistance, the voltage of which is proportional to the current flowing - and can thus be displayed in amps. The signal (typically 50 millivolts at 200 amps), is usually conveyed to an energy monitor via a light twisted pair lead that connects to the two small terminals on the shunt (such as shown here). Some now have a digital signal output.

*Figure 7.7. Typical current shunt. This one handles up to 200 amps.
Pic: rvbooks.com.au*

Current shunts that are combined with a main battery terminal are now also available (Figure 8.7)

If using a shunt with an existing regulator with an inbuilt monitor, buy the shunt from the regulator maker as it is likely to need a special adaptor that is supplied with that shunt. Installing a current shunt is not hard but unless you are familiar with dc electrical practice it is better to have an auto electrician do it for you.

Figure 8.7. Combined current shunt and battery terminal. Pic lem.com.

With a freestanding energy monitor, all current *must* flow through the shunt, so inputs and loads connect to its non-battery side. If used in conjunction with a solar regulator with inbuilt monitoring, the regulator solar feed must go directly to the battery, or the battery side of the shunt, otherwise solar input is shown (incorrectly) as doubled. The schematic drawing in Chapter 27 (Figure 2.27) shows a typical shunt installation. A 200 amp shunt is adequate for most RV use as it is not used to monitor the typical 500-600 amps starter motor draw.

Vehicles that have variable voltage alternators (Chapter 15) *must* have all negative returns taken to a common point on the vehicle chassis. This is because the earth strap from the chassis to battery (and/or alternator negative) doubles as a current shunt used to control alternator voltage. In some recent vehicles, however, that shunt is built into the negative battery terminal connector.

Figure 9.7. Crimp lugs and connectors are made in various forms and sizes. Pic: rvbooks.com.au

The alternator charge to the auxiliary battery can be included by taking the dc-dc charger's output to the battery via the shunt (and many such chargers require that shunt to be used anyway).

Crimp connectors

Crimp connectors are often the only way that 12/24 volt devices can be connected. High quality crimp connectors work well, but cheap ones loosen as they age and become a major cause of hard to locate electrical woes. Some crimp connectors are pushed onto associated lugs. Others (for joining cables) are held within plastic sleeves.

Figure 10.7. A high quality crimping tool is essential to make secure crimps. It is impossible to make reliable crimps without a specialised tool such as this. Pic: rvbooks.com.au

It is essential to use the correctly-sized crimp connectors. The smaller ones are colour-coded. Red is for 1.0 to 1.5 mm^2 cable, blue for 1.8 to 2.5 mm^2 and yellow for 4.0 mm^2 -6.0 mm^2 (Figure 9.7).

Crimp lugs are made in several hole sizes (but using the same colour codes). Crimp connectors made for larger diameter cables are not colour-coded.

Crimp connector quality varies - from cheap and nasty, to excellent. The best are of aircraft quality. Buy only from electrical suppliers and advise that you are seeking truly high quality units. Avoid the cheap auto parts and hardware store items. They are impossible to crimp correctly and cause problems after a few years' use.

Correct crimping

High leverage is needed to form the cold weld that is possible with top quality extruded lugs. Successful crimping needs the ratchet tool shown in Figure 10.7. It is impossible by using pliers or a vice.

Auto electricians have heavy crimping tools required for starter battery cables and will often crimp the connections for you.

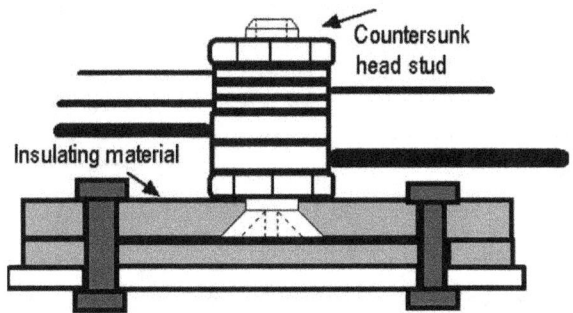

Figure 11.7. Power posts are readily made but also stocked by electrical suppliers. Pic: rvbooks.com.au.

The most commonly encountered crimping problems are corrosion, and wires working loose through faulty crimping. All introduce voltage drop, intermittent operation and eventually failure. Dirty connectors can be cleaned, but if corroded, fit new crimp connectors and protect then against dirt and mud.

Never solder an RV's cables. It initially provides good electrical and mechanical connection but, over time, corrosive flux penetrates cable strands. It also stiffens the cable locally - then fatigue failure sets in and the copper fractures.

Figure 12.7. This typical connector (cover removed) accepts one 25 mm² cable and three 16 mm² cables. Pic: rvbooks.com.au

Power posts/connector boxes

Avoid multiple leads hanging off the battery. Instead, run short heavy cables to power posts or connection boxes.

Power posts are threaded studs. They can be made from two layers of non-flammable insulating board and a countersunk screw held by a locking nut (Figure 11.7).

Earth return posts can be a threaded stud welded to a solid section of the chassis.

Also available are connector boxes - Figure 12.7. There are various configurations to hold up to 20 or more cables of sizes - from 1.5 mm² to 150 mm². Some accept one large main battery cable and multiple others of various sizes. Most connector boxes have red, black or clear covers that identify polarity. Clipsal (Australia) and Bulgin (UK) have good ranges suitable for 12 to 230 volts.

The wiring layout

Figure 13.7 below typifies that of basic auxiliary systems. Individual needs vary but most RV electrical systems will be much like this.

As shown, a circuit breaker can usefully double as a main switch providing it is readily accessible, yet close to the battery bank. Depending on distance, and the number of subsidiary circuit breakers, it is good practice to run individual leads from a main power post (or connector box) to each circuit breaker.

Cable runs should be as short as feasible, particularly those carrying heavy current because voltage drop across them is reflected throughout the system.

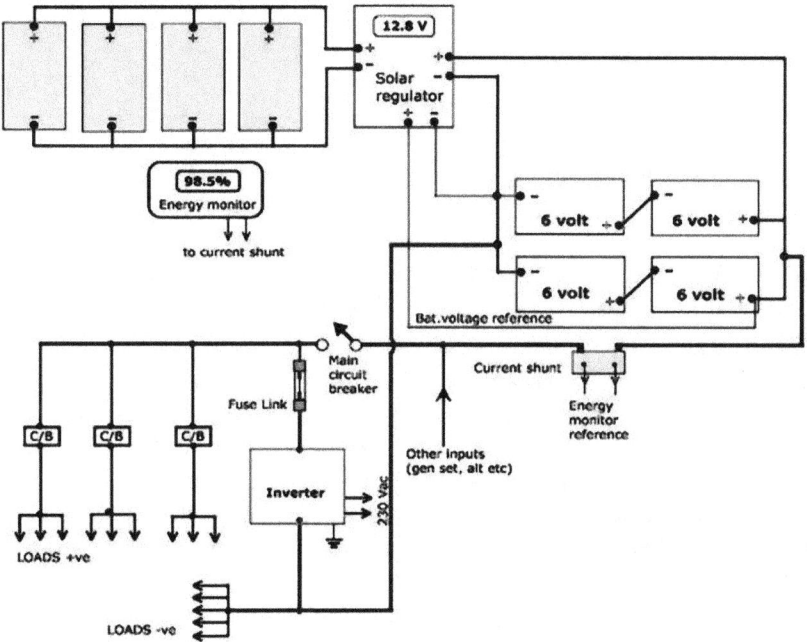

Figure 13.7. Most RV electrical systems will be much like this. Solar modules (top left) and battery banks (upper right) may be smaller or larger. That shown connected (of four series-parallel-connected 6 volt batteries) is common because single high-capacity 12 volt batteries can be too heavy to move safely in confined spaces.
Pic (©) 2018: rvbooks.com.au.

Chapter 8

Mains-voltage wiring

Recreational vehicles (RVs) are exposed to situations that require enhanced electrical protection. The requirements are set out in AS/NZS 3001:2008 as Amended in 2102 - a specialised extension of the main Standard (now) AS/NZS 3000:2018. There are minor differences between Australian and New Zealand requirements. These are made clear in both Standards.

In Australia all work above 50 volts ac on an electrical installation must be done by a licensed electrician. This still applies if the cabin or RV's 230 volt system is supplied only by an inverter or generator, even if there is no provision for external mains connection. The electrician is responsible for final inspection and testing. Some trades, however, are licensed to do limited and prescribed electrical work subject to final inspection by a licensed electrician.

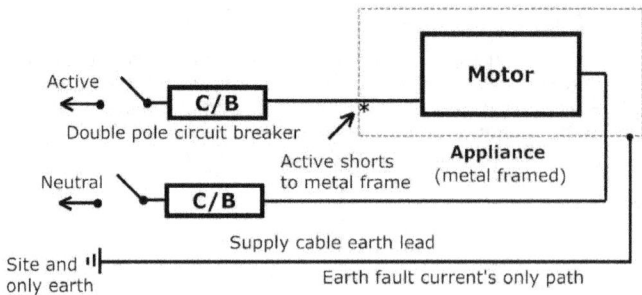

Figure 1.8. RV circuit breakers must be double-pole (i.e. they must switch both active and neutral). Pic: rvbooks.com.au

These regulations specify only minimum requirements. Electrical installation is very competitive so, unless otherwise agreed, only that minimum is typically provided.

There is a good case for exceeding some requirements. For example, it is legal (within limits) to run power and lighting from the same circuit but it is better practice have them separated, each protected by an individual CB (Circuit Breaker) Figure 1.8, and RCD (Residual Current Detector - Figure 2.8).

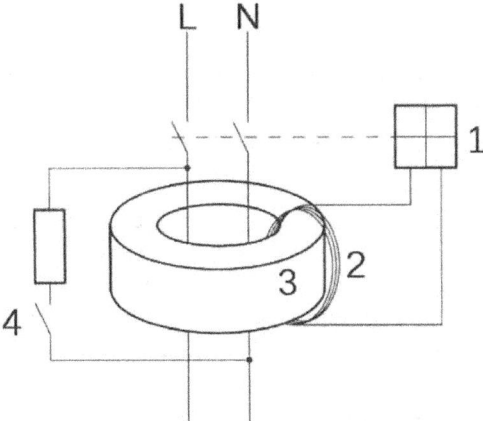

Figure 2.8. How an RDC works. Any difference in active (L) and neutral (N) current 'magnetises' the torodial core (3). This results in a current being created in coil (2) that causes the actuator coil (1) to cut incoming power.

If installing both 12 or 24 volt and low voltage cabling (e.g. 230 volt), they must be physically separated unless the 230 volt cable is double insulated, or 230 volt cable is used also for 12/24 volt cabling. It is simpler and safer to keep them apart. The reference is 3.9.8.3 in AS/NZS 3000:2018.

Electrical networks attract lightning strikes, static electricity, etc., that may damage equipment and are also potentially lethal to those using them at the time. To protect against the above and other hazards, electricity supply systems have 'neutral' supply cables linked to an earth cable. That cable is literally connected to earth.

How electrical earthing works

Our earth has 'zero voltage' and acts a 'sink' if, for example, struck by lightning. It likewise protects against electric shock from exposed conductive parts of electrical equipment keeping those parts close to earth potential if a failure of electrical insulation occurs. When a fault occurs, current flows from the power system to earth. This is ensured by connecting so-called 'neutral' conductor to an earth conductor.

An alternative to protective earthing, 'double insulation' ensures that a single failure or improbable combination of failures cannot result in contact between live circuits and any metal part that people may touch. Most domestic appliances are protected this way. Hand-held 230 volt power tools usually have extra electrical insulation between live components and anything metallic that the user could touch.

Functional earthing

In homes etc., the required earth-neutral link is normally made within the protected premises. With Australian, (and now New Zealand) RVs, however, that link is made within the electrical supply system via the supply cable.

Earth protection relies on soundness of the overall earth connection (that includes the supply cable). It is prejudiced by RV owners filing down a 15 amp plug's earth pin to fit into a 10 amp power socket. Further, earthing cannot protect against anyone accidentally contacting both active and neutral leads. See also 'Ten to 15 amp cable issues' - later in this Chapter.

RCD/CB protection

Extensive research, showing that healthy human hearts tolerate substantial current flow for about 0.5 second, resulted in a new approach that compares current flowing in the active and (earthed) neutral conductors. If unequal, current from the active lead can *only* have found an alternative and/or additional path to earth. To protect against this, a so-called residual current device (RCD) detects the unequal flow and cuts off the power within 0.4 second.

The RCD system became mandatory in new installations from 2000 onward. For it to operate reliably and within that 0.4 second, however, the electrical system *must* be installed correctly.

For further protection, an obligatory circuit breaker (CB) limits the current drawn. If that happens, in the event of live and neutral conductors touching, or someone attempting to run too heavy a load, etc., the circuit breaker likewise cuts off the power.

Cable Rating

Cable rating	Conductor area	Length
10 amp	1.0 sq mm	25 m
10 amp	2.5 sq mm	60 m
10 amp	4.0 sq.mm	100 m
15/16 amp	1.5 sq.mm	25 m
15/16 amp	2.5 sq.mm	40 m
15/16 amp	4.0 sq mm	65 m

Table 1.8. This (Table 5.1 from AS/NZS 3001:2018), lists legally available supply cable ratings and lengths.

To ensure the contact breaker acts in time, all associated 230 volt ac cables must be of the prescribed size and maximum length that ensures the voltage drop does not exceed 5% under full load. Such sizing ensures current flow does not extend beyond that length of time required to protect life.

Earthing is still essential.

Used correctly, the above protects cabling and appliances against failure or damage. This, however, tends to be prejudiced by ongoing campfire and forum misinformation.

To ensure circuit breakers (CBs) work effectively, the supply cable *must* have sufficiently low impedance (i.e. resistance to ac current), for the circuit breaker to cut current flow within 0.4 second. The latter is prejudiced by voltage drop over 5%, hence the longer the cable and/or the higher its rated current, the thicker must be that cable.

All RV supply cables must thus accord with that specified in Table 5.1 of AS/NZS 3001:2008 as Amended in 2012 and AS/NZS 3001:2018

Never join supply cables together

Joining supply cables together to extend overall length is seriously dangerous. The now longer length may limit current flow and preclude the circuit breaker tripping in time to save life.

Caravan park supplies

Caravan parks are legally required to have CB/RCD protected three-pin 15 amp outlet-sockets, but a few still have only 10 amp outlet sockets. Many homes too have only 10 amp outlet sockets. Most RVs, however, have a 15 amp inlet plug and 15 amp circuit breaker. A way of coping with 10 amp 230 volt outlets is shown below.

Ten to 15 amp cable issues

Because a 15 amp plug earth pin is bigger than a 10 amp plug earth pin it will not fit into a 10 amp socket. This presents problems where only a 10 amp supply is available. Some owners make an illegal 10 to 15 amp adaptor cable or file the 15 amp plug's earth pin to fit. They place themselves at risk and face serious legal charges in the event of causal harm or death.

The Ampfibian adaptor (Figure 3.8) is a good solution. It is a 1.5 metre inlet cable device that has an inbuilt circuit breaker that restricts supply current to 10 amps. It also has an RCD. Its 15 amp socket accepts an RVs' 15 amp plug and cable. (It is legally classified as a 'Portable Socket Outlet Assembly'.)

Figure 3.8. The Ampfibian enables a 10 amp cable to be used if no 15 amps outlet is available. Pic: Ampfibian.

If 15 amps is *never* required, one can legally use a 10 amp cable and 10 amp socket inlet (also enabling a 100 metre cable if needed). The standard RV inlet socket is 15 amps, but 10 amp RV inlet sockets are now available.

Another legally acceptable alternative is to have a supply cable sized as in the Table 1.8 (in this Chapter) *permanently* connected to the RV. The cable must be securely anchored, and storage provided. If a 10 amp cable and plug are used, the existing 15 amp circuit breaker/s and RCD *must* be replaced by 10 amp equivalents. The 15 amp switched socket outlets in the RV may be retained if wished.

Polarity explained

If a plug, socket or supply cable has reversed polarity (i.e. active and neutral lead crossed over) any connected appliance is electrically alive but does not work until switched on. Users have no warning they are at risk. Domestic systems are installed by qualified electricians who check every outlet to ensure polarity is correct.

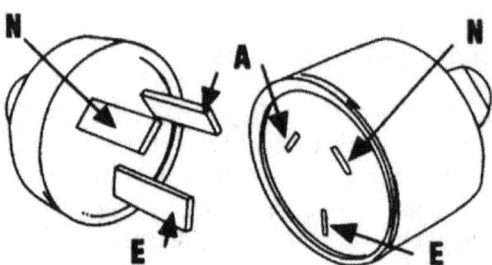

Figure 4.8. Correct connections for three-pin plugs and sockets.

Certified appliances have long been supplied with a fitted supply cable and (now non-openable) plug. With these, there is almost zero risk of active/neutral leads being reversed, or an earth connection omitted. This being so, switches in domestic wiring need to break only the active conductor. It is, however, not unknown for caravan park outlets to have incorrect polarity.

Reversed polarity in an RV is more likely because people illegally make up their own supply cables. There is thus a risk of accidentally reversed conductors at the supply cable's plug or socket. Figure 4.8 shows the correct connections.

Polarity checking

Polarity is readily checked by using a plug-in tester (Figure 5.8). The tester is plugged into the outlet socket and the power switched on. Coloured LEDs indicate if polarity is correct or otherwise. Most also test for a missing or inadequate earth.

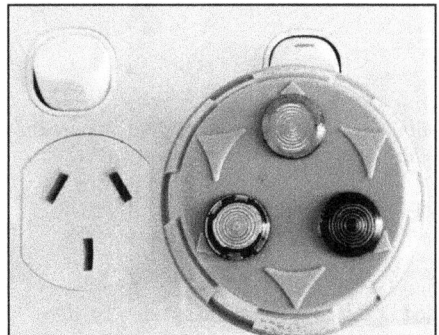

Figure 5.8. Polarity testers typically have red, green and amber lights that indicate correct or incorrect polarity. Pic: rvbooks.com.au

Polarity testers do not directly pin-point the wiring error's location. That is done by working backward, i.e. toward the 230 volt supply outlet. The cause is usually an incorrectly wired supply cable plug or outlet socket.

Double-pole switching

To safeguard RV users against the only too real risk of incorrect polarity, all RVs (in Australia) must have their outlet socket protected by so-called double-pole switches and circuit breakers. These open and close both active *and* neutral conductors - Figure 6.8.

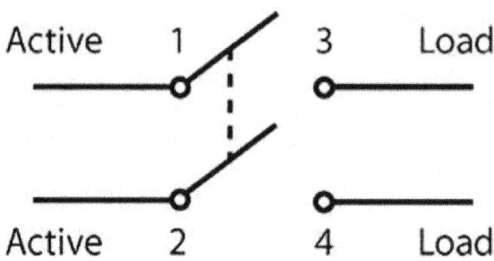

Figure 6.8. Double-pole switching - both active and neutral are switched. Pic: rvbooks.com.au

Cable sizing & installation

Depending on load, 230 volt cabling within RVs must be multi-strand 1.5 mm² or 2.5 mm², not the heavy single strand type used in houses. This is because ongoing movement causes single strand conductors to fatigue and break.

Mains wiring and 12/24 volt wiring must be physically separated, or the mains cable must be double insulated. It is preferable to install 230 volt and 12/24 volt cables well apart. If 230 volt lighting cable is covered by heat insulation, 2.5 mm² cable must be used.

All RV 230 volt cabling that is at risk of damage or disturbance must be run within conduit. Elsewhere it may be run along battens, etc., and supported at not more than 300 mm intervals or, if in an enclosed space and resting on a horizontal surface, at not more than 500 mm intervals. Wherever it passes through metal RV 230 volt cabling must be protected by insulating grommets It is also good practice to chamfer the holes where the cable passes through timber.

Inverters

Inverters convert a typical 12 or 24 volts dc to 230 volts ac. Those that have socket outlets *on the actual unit* must not be connected to any *fixed* 230 volt wiring. Appliances may only be plugged into the inverter (Figure 7.8).

Figure 7.8. Inverters that have inbuilt outlet sockets, such as this Projecta unit, must not be connected to any mains fixed wiring. Pic: Projecta.

Inverters intended for connection to fixed 230 volt wiring *have* no inbuilt sockets. They *must* be installed by a licensed electrician.

Unless you totally know what you are doing, use only inverters that produce a true sine-wave and are known to be double insulated. Those that are will usually be clearly promoted as being double insulated. See also Chapter 16.

Generators

If a generator feeds an RV via its 230 volt inlet, it must meet the requirements of AS/NZS 3010. Details (applicable also for inverters) are shown in AS/NZS 3001:2008 as Amended in 2012.

Generators that do not meet AS/NZS 3010 (those that do have a rating plate that advises accordingly) may only be used by plugging appliances directly into their outlet sockets. They absolutely must not be connected to fixed wiring. Those that do meet AS/NZS 3010 must be connected to fixed wiring as shown in AS/NZS 3001:2008 as Amended in 2012.

Change-over switches

All mains-connectable systems that have an inverter or generator must have a double-pole 'break before make' changeover switch (Figure 8.8).

Figure 8.8. Change-over switch has a 'break before make' action. Pic: Clipsal.

This switch ensures that the inverter or generator cannot accidentally feed power back into the network, even momentarily. It also protects electricians working on wiring that is assumed not to be alive. Such protection is built into mains-connect inverters.

Lights & appliances

Regulations govern type and placing of 230 volt electrical fittings and outlets near sinks, showers and water generally. These regulations too are covered in AS/NZS 3000:2018.

Disclaimer (professional)

Information in this section is included only as a guide to what should be. It is not intended as an inducement for those not qualified to undertake such work.

The coverage is therefore limited to what RV owners need in order to have some concept of what is required, and also of the reasoning behind the requirements. It may assist to detect anything obviously wrong. It may also assist knowing if a prospective purchase is likely not to meet required Standards. Specialised knowledge is required to be sure that installation meets all requirements for the now often required electrical re-certification.

This section is not in any way intended as a substitute for having an experienced licensed electrician make a thorough assessment and provide a definitive written opinion.

All content section has been checked against the latest AS/NZS Standards. Totally new Standards are rare but AS/NZS 3000:2007 as Amended in 2012 was updated to AS/NZS 3000:2018 in mid-2018. AS/NZS 3001:2008 as Amended in 2012 is still current at the time of publishing this edition.

These (very costly) Standards used to be available via public libraries but, for reasons unclear, they are currently no longer so available.

Chapter 9

Batteries (general)

Regardless of their type (conventional wet cell, sealed, Amalgamated Glass Mat (AGM) and the now rarely used gel cells) all lead-acid batteries work in much the same way. They store energy via chemical interactions between lead plates and electrolyte (dilute acid). The amount of energy stored relates to the electrolyte's concentration.

Chemical interactions are initiated by applying a voltage across the battery (charging) or via a load across the battery (discharging). The interaction is far from instant but the larger the plates' surface area, the quicker. Starter batteries must react quickly, speed is less necessary for RV auxiliary use.

Batteries are designed and made for various purposes and selected and used accordingly. Some lead acid batteries are made specifically for engine starting, other types of lead battery (often known as 'deep cycle') power the RVs accessories. A few can be used for either purpose. Most lithium batteries are multi-purpose.

Construction of a lead-acid battery

A lead-acid battery typically has one to six cells, each of pairs of positive plates of lead dioxide and negative plates of high surface-area porous sponge lead. These plates are held apart by inert plastic or glass-fibre separators.

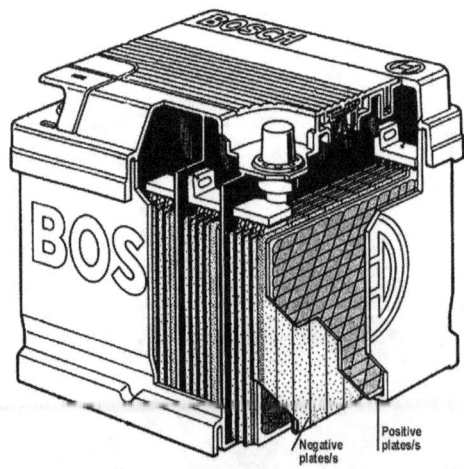

Figure 1.9. A conventional deep-cycle battery has a small number of thick plates and heavy connecting links between plates and cells. Pic: Bosch.

The older lead acid 'wet cells' were filled with a dilution of water and sulphuric acid that covered their plates.

AGMs have electrolyte held within a fibreglass matrix; gel cells within a candle-like substance.

Each cell stores energy at a little over 2.1 volts when fully-charged. Cell volume determines how much energy each can store and release, but does not affect its voltage.

Lead-acid batteries marginally increase in capacity during their first few months of their working lives. New batteries may thus have that marginally less capacity than claimed. Once maximum capacity is reached, it then falls at a rate related to usage rather than age.

Starter batteries

Contrary to general belief, starting a big engine (still mostly via a lead-acid battery) requires surprisingly little energy. Today's big 4WD's diesel engine starter motors draw about 500 amps but rarely for more than two or three seconds. Even allowing for major inefficiencies this is only about that amount of energy drawn by a five watt LED in one hour. It depletes the starter battery by only 2% or so, and is replaced within two to three minutes by the alternator.

Figure 2.9. Hand cranking a car required care but not a great deal of strength. Pic: Henry Ford Museum.

A starter battery's ability to supply high short-term current requires it to have multiple thin plates, with large surface area. This assists speedy chemical interaction with the electrolyte, enabling heavy current to be released for brief periods. It equally enables that energy to be rapidly replaced.

A starter battery's multiple plates have a large surface-area that results in ongoing internal leakage. Unless the vehicle is used regularly, the batteries lose about 10% or so charge a month. This causes active lead material to be shed from the plates and to build up from the cell's bottom. If that build-up reaches the plates, it short circuits them, resulting in instant failure.

Standard starter batteries are a compromise between size and performance. Most last three to four years in mild climates. To avoid being caught out, many who travel extensively in outback areas replace starter batteries routinely every second year. How to cope with vehicles unused for long periods is shown in Chapter 11.

Cold-cranking amps (CCA)

Starter battery output is reduced at low temperature but a starter battery must produce current in the coldest conditions likely to be experienced. This is reflected in their associated standards.

The US standard, used also in Australia, defines the number of amps the battery must sustain while cranking for 30 seconds - with voltage staying above 1.2 volts per cell (corresponding to 7.4 volts for a 12 volt battery) at -17.8°C. The British standard maintains cranking for 180 seconds, to 1.4 volts per cell (8.4 volts), at -17.8°C. The International Electrotechnical Commission maintains cranking for 60 seconds, down to 1.4 volts a cell, also at -17.8°C (i.e. about 0°F).

The Marine Cranking Amps (MCA) standard is similar to the US standard but measured at 0°C. This results in higher numbers that can mislead if it is not realised that it is the much higher temperature at which it is measured that enables the longer maintained cranking.

In warmish countries (and using the US standard), a typical petrol engine needs about 80 to 90 CCA per litre. Because of its high compression ratio, a typical diesel needs about twice that.

Reserve Capacity

This is an industry-defined measure of the length of time that a fully-charged starter battery can perform if discharged at 25 amps before dropping below 10.5 volts (for a 12 volt battery).

It is often claimed that a battery with high CCA has high reserve capacity, but this is not necessarily so. Increasing CCA requires more and thinner plates but that reduces reactive material and hence Reserve Capacity. The latter capacity is primarily related to battery size and weight.

Beware of normal-sized batteries with exceptionally high CCAs - as reserve capacity may be low. As a rough guide, reserve capacity can be converted into Ah by multiplying by 0.4. Buying a physically large starter battery that has the CCA you need usually guarantees ample reserve capacity.

Lead/calcium batteries

Mainly starter batteries, these use lead/calcium plates to reduce gassing and water loss. They have more space above the plates and charge at higher voltage.

Deep-cycle batteries

The term 'deep-cycle' battery may mislead. These batteries have a small number of thick plates enabling them to withstand repeated discharges, but only to about 50% or their life is unduly shortened. They cannot provide high current because chemical reactions within the thick plates are too slow to maintain cell voltage. They are damaged if that is done frequently.

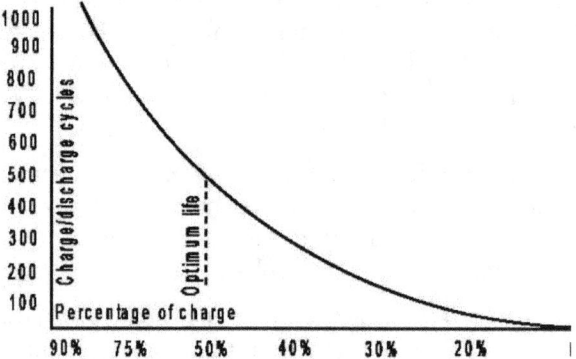

Figure 3.9. Deep-cycle battery life versus depth of discharge.

The most common measurement of capacity is amp hours (Ah) but the ability to release energy (i.e. power) from this type of battery is dependent on the *rate* of discharge. Because of this, deep-cycle battery makers specify the battery's usable capacity in those terms. That mostly commonly used is the '20 hour rate' (also known as the '5% rate'). Some makers use other rates, so check that if comparing respective capacities.

At the 20 hour rate, a fully-charged 100 Ah battery may sustain a current of 5.0 amps for 20 hours (100 Ah) while staying above about 10.5 volts for a 12 volt battery. If discharged at 50 amps however, that same battery can typically sustain that draw for only one hour.

Contrary to common belief the remaining 50 amp hours are still available - but at a lower voltage and lower rate of discharge. Supplying a microwave oven's typical 130 amps with such batteries of less than about 300 Ah substantially shortens their life. Unless of 300 Ah, use AGM or lithium batteries.

Marine batteries

These are part starter battery and part deep-cycle battery. They are intended to provide energy storage, plus occasionally to start a small engine in a boat. They are often touted as being 'superior' (they're marine quality, mate - they have to be better!). In reality they are just different. They have the high-current failings of non-marine conventional batteries but to a lesser extent. There's little point in using them in cabins or RVs.

Sealed batteries

RV starter motor and the RV's auxiliary batteries (that power the domestic bits) have long since been sealed - but have vents that relieve pressure in the event of a major fault. Their construction, of both starter and deep-cycle types, is generally similar except for changes to plate material. There is also a larger volume of electrolyte that, theoretically, is used up at about the same time that the batteries are due to expire.

These batteries outlast those that need regularly topping up with water but do not match those (rare few) that are properly maintained.

Gel cell batteries

Unlike conventional lead-acid batteries, gel cells have their electrolyte, of sulphuric acid and phosphoric acid (that gelled by fumed silica or pumice), hardens like candle wax. It then forms cracks and fissures between the positive and negative plates.

During charging, oxygen from the positive plate trickles through to the negative plate via those pathways. During discharging, sulphuric acid migrates in a similar but reverse manner: phosphoric acid migrates to the electrolyte, ensuring the electrolyte remains conductive.

These batteries build up internal pressure that is critical to their operation, resulting in minor case bulging that may concern if not realised that it is normal. Automatic vents open in the event that internal pressure exceeds dangerous levels.

As with all 'sealed batteries', gel cells must not be used in enclosed areas small enough to enable the concentration of hydrogen gas to build up to the level that can result in an explosion.

Gel cell batteries have low self-discharge that is heat dependent. In temperate climates the loss is under 1% a month but increases rapidly above 40°C. In such temperatures they need recharging every three months. Their characteristics are in some ways similar to marine batteries in that they can provide higher current than a deep-cycle battery. Whilst still made they are increasingly challenged by the AGM batteries described below.

Absorbed glass mat batteries

Known generally as AGMs, these were initially developed for military use in Arctic regions. They were in general use by 1990.

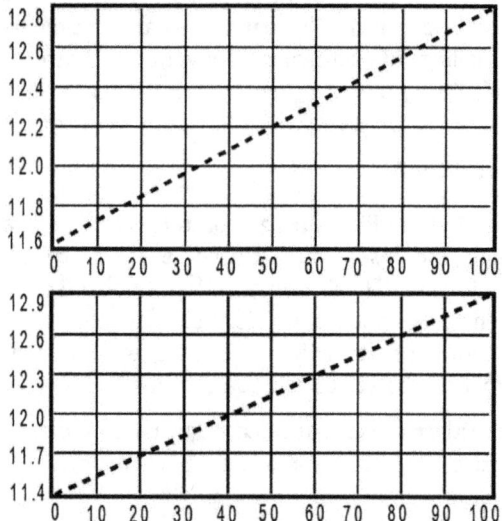

Figure 4.9. Off-load voltage and percentage discharge for (top) conventional lead-acid batteries and (lower) for AGM and gel cell batteries.

There are minor differences between AGM battery types but all have multiple plates with thick glass-fibre mats that hold the electrolyte in suspension. This aids mechanical strength so they are well-suited to 4WDs and off-road camper trailers.

A few AGM batteries are marketed as suitable for starter motor and deep-cycle use but any AGM battery bank of 250 Ah or more is able to start a 4WD engine if the starter battery fails.

Their achievable charge/discharge cycles are fewer than from well-maintained conventional deep-cycle batteries but, as few of the latter are so treated, most users find that AGM batteries last longer. They are more readily charged, and can be repeatedly discharged down to 30% (Figure 4.9). They thus have more usable capacity than conventional lead-acid batteries.

An AGM's minor downside is its greater weight: a 12 volt 100 amp hour AGM weighs about 33 kg. They also cost about 50% more than their lead-acid deep-cycle equivalents.

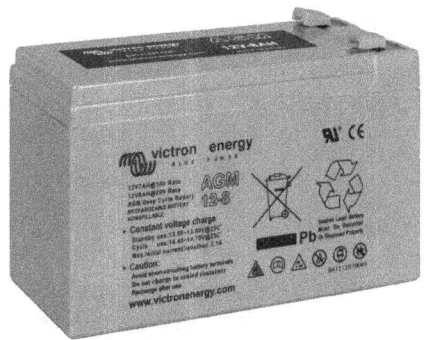

Figure 5.9. AGM battery: Pic Victron.

AGM batteries dislike heat and will not last long under a vehicle's bonnet, especially if subjected to the considerable heat from a turbo working hard.

On the plus side AGM batteries hold their charge for six to twelve months and are mechanically rugged. They are a very good choice for coach conversions where their size and weight is less of an issue.

See Chapter 11 regarding issues relating to float charging (a process used with conventional lead acid batteries when unused for long periods) but not AGMs.

Lithium batteries

There are various types of lithium batteries. Some early examples (of very high energy) were prone to thermal runaway, resulting in fires. Those used in RVs (LiFePO4) use the less energy-dense combination of a lithium cathode, and carbon or graphite anode, plus electrolyte.

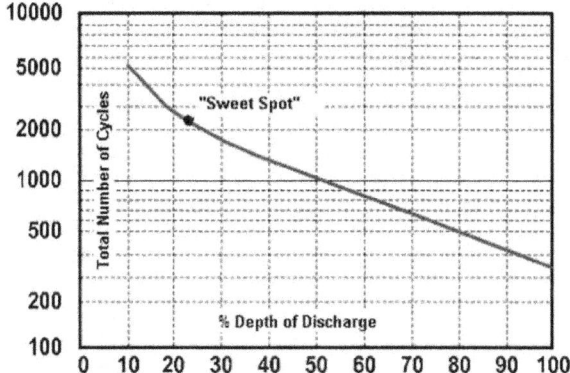

Figure 6.9. AGM batteries withstand repeated deep discharges very much better than conventional batteries. Even if 90% discharged they will still survive about 400 such cycles.

LiFePO4 batteries store about three times more energy than other batteries of similar size and weight. They can be deeply and routinely charged and discharged without harm.

These batteries cost more for similar capacity, partially offset by more of that capacity being routinely available. They also maintain an almost constant discharge voltage (13.1 to 12.9) regardless of (typical RV) loads.

In 2019, their sales are still mainly limited to specialist suppliers.

A few vendors claim that LiFePO4 batteries are direct replacements for existing batteries. Be wary of such claims as LiFePO4's have specific charging needs. Current issues will be overcome but there are evolving battery technologies that may well have similar appeal with fewer issues (actual or perceived).

Battery charge & life

With any rechargable battery, available capacity is limited by the level to which it is charged and routinely discharged. A conventional 100 Ah battery charged to a realistic 90% and discharged to the typically recommended 50% provides only 40 Ah. (Voltage versus % charge is covered in Chapter 11).

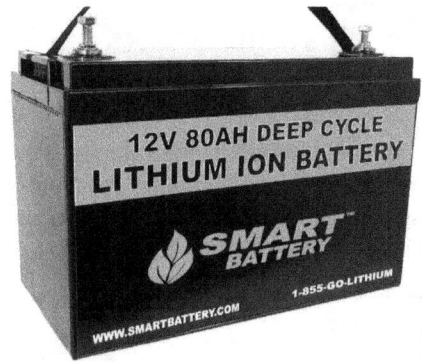

Figure 7.9. Lithium battery. Pic: Smart Battery.

Used this way, a good quality deep-cycle battery in a temperate climate will provide about 500 cycles of such use. In practice, however, many RV users routinely discharge batteries until the lights dim and the fridge stops cooling. This corresponds to about 11.4 volts (i.e. little remaining charge) and slashes conventional battery life. There is also risk of such usage damaging your appliances. A water pump relies on the water pumped through it for cooling. It may stall and burn out if voltage is too low. Many refrigerators shut down when the battery is deemed to be about 80% discharged.

For solar, conventional batteries should ideally be sized such that they discharge no more than 15% overnight (85% remaining) and fully charge by noon on most days. This enables occasional discharges to 50% or so without overly shortening their life, providing they are fully charged immediately afterwards. In such use, a good quality deep-cycle battery typically copes with 1000 or so charge/discharge cycles.

Conventional battery life relates also to the charging regime - doing this correctly extends its life. Assessing battery charge and discharge rates is covered in detail in Chapter 11. In essence, such battery reality is that one buys amp hours, that are then used as desired.

LiFePO4 battery life-span is claimed to be partially dependent on their percentage of full-charge: 80% to 90% being increasingly quoted as the limit for 2000 cycles. Those in cabin and RV use also maintain almost constant voltage for 90% to 20% of their charge.

Capacity - how much?

Except for large truck and coach conversions (and LiFePO4 batteries), RV battery capacity is usually a compromise between the ability to accommodate its bulk and weight, and the ability to fully recharge it each day. As a generalisation, maximise charging ability, not battery capacity. Unless the former is available, the energy is not there to store.

Self-discharge

Conventional lead-acid batteries self-discharge at a rate dependent on their type, condition and temperature. Deep-cycle batteries in reasonable condition lose 2% to 3% per week during winter, and about twice that in summer. In temperate climates, AGMs lose very little. If stored they may then need only minor recharging once a year. Chapter 11 shows how to cope.

Lithium batteries can be left, for years if necessary, almost regardless of their state of charge. In practice, most needing to do so (e.g. military users) leave them about 50% charged. Their internal self-discharge is far too low to be of any concern - even for years.

What type of battery to choose

Ongoing deep discharges decreases conventional lead-acid batteries' life. Gel cells and AGMs are less affected but are large, heavy and costly; both charge quicker and deeper, particularly via dc-dc charging.

Gel cells and AGMs will not last as long as conventional lead-acid batteries that are optimally maintained, charged and discharged - but few are. Lithium appeals for RVs as most have limited space and weight carrying ability. At present (mid 2019) not all battery suppliers stock them.

LiFePO4 batteries are recommended but caution is needed unless you understand this technology or know someone (or a company) that does. For ongoing updates please see: rvbooks.com.au

Chapter 10

Installing batteries

All batteries must be treated with respect, they can be very dangerous or even explode if misused. A spanner dropped across battery terminals, or a live battery cable accidentally earthed, causes thousands of amps to flow though that spanner. It may even vapourise the spanner. This is particularly do with lithium batteries regardless of size. If you are working on or near a battery or thick live cables, wear a face shield, or safety goggles plus heavy overalls. Always first remove your watch and any metal jewellery.

All batteries must be housed in a cool, dry, ventilated enclosure, and the tops and terminals kept clean and free of corrosion. Fit strong rubber or plastic insulating caps over otherwise-exposed terminals.

Ventilation is still essential

Despite battery makers warning that ventilation is still essential, many RV makers choose to ignore this and locate those batteries in non-ventilated areas.

Beyond 70% or so charge, lead-acid batteries begin to produce highly flammable and potentially explosive gas. They typically produce 0.42 litres of hydrogen and 0.21 litres of oxygen per amp hour.

Most batteries are now sealed but must withstand considerable internal pressure. To avoid risk of an explosion in the event (say) of a charging fault causing gross overcharging, they have safety vents that release hydrogen (at high pressure).

The MK Battery company warns that these so-called sealed valve-regulated batteries are very safe unless abused. It states that 'However, as with any type of battery, certain safety precautions must be taken ... Do not install any lead-acid battery in a sealed container or enclosure. Hydrogen gas from overcharging must be allowed to escape.'

Emitted gas burns with a fizzle at a concentration (in air) of around 4%. At 14%-18%, a resultant explosion can blow a structure apart. It requires a tiny amount of energy to ignite - about 0.02 joule (far less than the heat of a candle). That can be caused by poorly-secured terminal clamps and cables, battery connectors that work harden and crack and tiny sparks from worn bearings.

Never install switches, solenoids, battery chargers, etc., in a poorly ventilated battery enclosure.

How much ventilation needed

There are no legal standards in Australia or New Zealand for the design or ventilation of battery enclosures in trailers or motorhomes. The Australian Clean Energy Council, however, advises the following vent area for fixed installations:

Area = 0.006 x C x I - where: area equals minimum of each vent in cm², C equals number of cells, and I equals maximum charging rate (amps).

For RVs it pays to double this vent area. It is still not large.

The vents need to be at the very top and bottom of the enclosure and diagonally located. (Industry practice is to have a few 25 mm diameter holes at the top and bottom of the enclosure).

Series or parallel?

Large batteries (e.g. above 100 Ah) are heavy and unwieldy. It is easier to move several of medium-size. Batteries can be connected to obtain higher voltage or higher current (Chapters 9 & 10) but one cannot obtain energy for nothing. Any combination of the same batteries results in *exactly* the same amount of energy being stored.

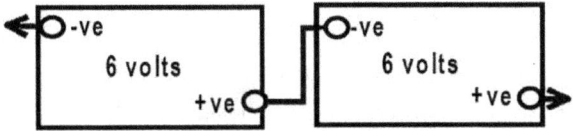

Figure 1.10. Series connection: each battery is 6 volt, 100 amp-hour. Output is 12 volts at 100 amp-hour. Pic: rvbooks.com.au

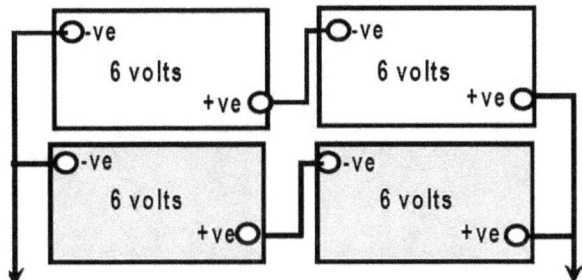

Figure 2.10. Series-parallel connection: each battery is 6 volt, 100 amp-hour. Adding the shaded pair results in 12 volts at 200 amp hours. Pic: rvbooks.com.au

Connecting two 6 volt 100 Ah batteries in series, i.e, end-to-end (Figure 1.10) provides 12 volts at 100 Ah (1200 Wh).

Connecting the same batteries in parallel results in 6 volts at 200 Ah (still 1200 Wh).

Adding a second parallel pair in series-parallel (Figure 2.10) results in 12 volts at 200 Ah (a now 2400 Wh).

Many RV owners whose batteries discharge sooner than expected make matters worse by increasing battery capacity (thus increasing charging/discharging losses).

To increase the amount of energy stored you do not just need more, larger or different batteries but (and essentially) increased ability to charge them.

Which is better?

Within reason, series and parallel connection each have their own (mutually exclusive) pros and cons. According to the Exide Corporation, it is fine to parallel up to ten batteries of identical type, voltage, age and condition. They can have totally different capacities and each takes and delivers in accordance with its ability and needs.

Do not parallel-connect any other type of battery directly across AGMs or LiFePO4s as their higher charge acceptance causes them to grab most of the available charge. Instead, connect them via a voltage sensitive relay.

Series-connected batteries are limited by the weakest battery in the chain.

12 volts from 24 volts

A few converted trucks and coaches have 24 volt alternators and twin series-connected 12 volt batteries (and thus 24 volts), yet 12 volts is needed for lights and appliances. This is feasible by using either a charge voltage equaliser (Figure 3.10) or a voltage converter) Figure 4.10.

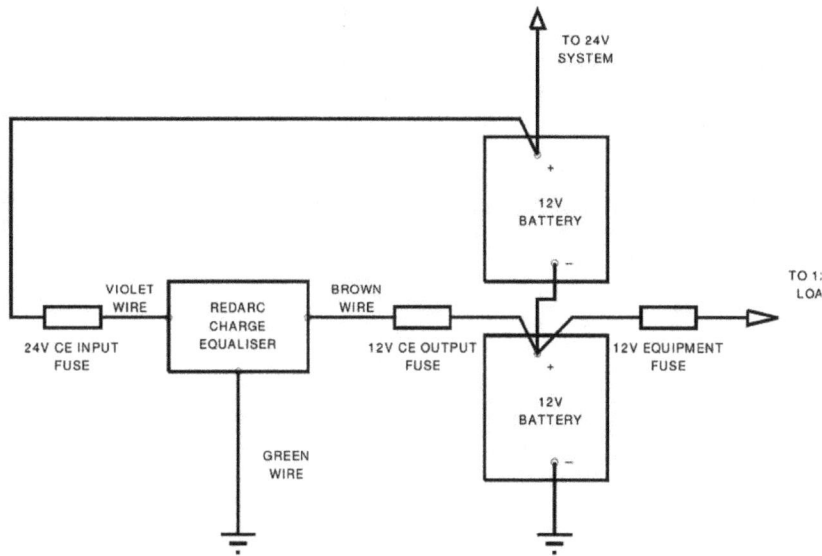

Figure 3.10. A charge voltage equaliser enables a heavy 12 volt load to be drawn from the upper battery. Pic: Redarc.

Voltage equalisers

Voltage equalisers enable 12 volts to be drawn from one of the two 12 volt batteries, whilst ensuring that both remain equally charged. They are not as efficient as converters but can supply full available battery current. They are made in varying capacities that relate to the equalised current, not the supply current.

Voltage Converters

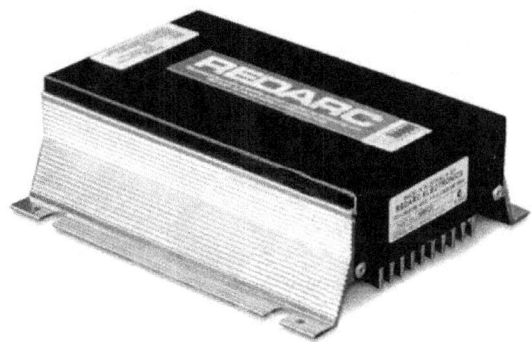

Figure 4.10. SMF30 converter provides a nominal 12 volts at up to 33 amps from 24 volts. Pic: Redarc.

Voltage converters change 24 volts to 12 volts electronically. They are more efficient than equalisers but limited in the maximum current they can supply.

The choice between converters and equalisers is usually obvious. If you need over 50-60 amps, e.g. for a microwave oven, you have no choice but to use an equaliser.

If it is borderline it is probably best to go for the equaliser as that can, if later found necessary, supply *much* higher current.

Chapter 11

Battery charging (general)

The dry batteries we use in torches, etc., contains energy, that once used up, are replaced by new ones. The type of batteries discussed here are rechargable. This is done by applying a voltage across a battery that exceeds that which it currently has. The *higher* that voltage difference, the *higher* the charging current, and the faster that battery charges. There are, however, upper limits that, if exceeded, will wreck the battery. The limits vary from one battery type to another. Because of this, battery chargers are programmable to suit battery type and capacity.

Until 2000, vehicle charging systems generated a nominally fixed voltage (e.g. 14.2-14.4 volts). If used to charge a lead acid or AGM battery, the *difference* between that fixed voltage constantly falls, so charging current likewise falls.

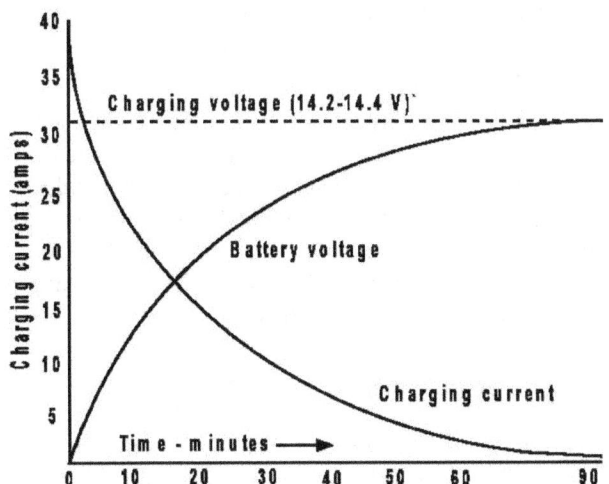

Figure 1.11. Typical vehicle charging system for a lead-acid battery. As the charging battery's voltage rises towards the alternator's output voltage, the charging rate rapidly falls. Pic: rvbooks.com.au

As Figure 1.11 shows, by 75% or so (of full-charge), charge current becomes a mere dribble but, if left on charge for a week or more, the battery may eventually overcharge. Many older RVs, and present-day cheap battery chargers still operate like this.

This charging compromise is not a problem for engine starting - the starter motor is designed accordingly. It is however, very much a problem when charging auxiliary batteries as these are similarly limited: they rarely exceed

75% charge, and less if located in a caravan - where 65% charge is still far from uncommon.

AGM and gel cell batteries have greater charge acceptance. They can cope with a constant voltage above 14.1 volts but charge faster and more fully using multi-stage charging. This is also the basis of the dc-dc alternator charging described in Chapter 13. That form of charging (also used in 230 volt battery chargers) is now the *only* method worth considering.

Multiple power point tracking (MPPT)

Multiple power point tracking is a technique that accepts a range of input voltage (within specified limits) and automatically and constantly juggles available voltage and current to optimise that which best matches the application's needs.

MPPT cannot increase energy as such but partially reduces otherwise-inevitable power mismatch losses. Initially developed for solar powered water pumping, MPPT is now built into many solar regulators, and dc-dc alternator chargers. Those MPPT units intended for use with small solar systems typically accept 9.0 to 36 volts and can be adjusted for 12, 24 and 48 volt batteries, and also for the optimal charging regime for the type and capacity of battery used.

With solar applications, MPPT recovers losses by 10%-15% or so but not the 30% plus many vendors claim. It also enables solar modules to be series-connected such that, for the same wattage, they produce higher voltage at less current. This also enables smaller cabling to be used from solar array to solar regulator. For alternator charging MPPT enables incoming voltage/current to be automatically 'tailored' to the regime that is optimal for the type of battery to be charged.

Multi-stage charging

Lead-acid batteries can be safely charged at any rate that does not cause them to exceed 50°C. Most high quality chargers follow a regime like that shown in Figure 2.11 and described below:

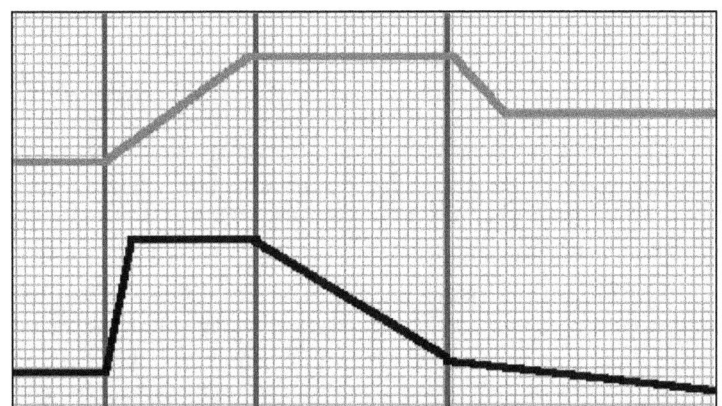

Figure 2.11. Basic multi-stage charging. Variants including a lower voltage warm-up stage and an (optional) equalisation stage prior to floating.

Initial bulk charge: This is typically an initial constant current of about 15% to 25% of the battery's nominal capacity. It is achieved by constantly increasing the charging voltage as the battery voltage rises until 14.2 - 14.4 volts - depending on battery type.

Absorption: charging is typically held at a *constant* 14.2-14.4 volt and the charging current progressively drops to about 10% of the battery's nominal capacity. The charging voltage is then either held at that, or dropped slightly for two to three hours to enable the plates and electrolyte to accept and evenly absorb charge. The rate of charge gradually falls as the charge becomes absorbed, eventually to 1% to 2% of the battery's capacity.

Floating: following absorption the charge is reduced to 13.6 to 13.8 volts. The battery slowly continues to gain charge and may reach 100%. In practice, 95% is most likely to be achieved with basic lead-acid batteries. AGMs and gel cells, however, accept charge more readily and 100% is common.

Equalising: now rarely used (and never with AGM or LiFePO4), is an optional stage intended to breakdown undesirable plate deposits (sulphation), and to ensure that interconnected cells and batteries accept a more equal charge.

Equalising applies a voltage high enough to cause a fully-charged battery to accept a further 5% of its amp hour capacity. It is held in this condition for two to three hours, during which it gasses strongly. As this dissolves plate lead it reduces battery capacity. Mainly used now only with the increasingly rare wet deep-cycle batteries, equalising is now an option with some battery chargers and solar regulators. Many battery makers specifically recommend against using it, or confining it to series-connected batteries, as these tend to develop unequal charge over time, but that is better corrected by disconnecting and charging them individually.

Overcharging

While rare, long-term overcharging can happen. It is usually caused by leaving a battery connected too long across a cheap charger, or by a faulty or incorrectly programmed solar regulator, or charging from a solar module without using a solar regulator. It also occurs with under-bonnet located batteries because the higher their temperature, the more readily they absorb charge (and overcharge).

To counteract this, some voltage regulators and chargers monitor battery temperature at the battery itself, and adjust charging voltage accordingly. This is essential for AGMs or gel cells in an engine bay (but as they are damaged by heat this is not a good place for them anyway). Some alternators (post 2000) reduce voltage to levels that are too low for effective charging. See Chapter 14.

Undercharging/over-discharging

As a conventional lead-acid battery discharges, sulphur combines with the lead, forming lead sulphate. If the battery remains uncharged even briefly (or is consistently undercharged) the sulphate crystallises and drops to the bottom of the cell. A starter battery's life is typically terminated by the shredded material piling up until it reaches and short-circuits the plates, usually killing it within days, and sometimes instantly.

Deep-cycle batteries usually shed material progressively until there's not enough left to do anything useful. Sudden failure is rare but capacity is progressively lost. Eventually they are unable to hold a charge for more than a few days. It also manifests as a higher than expected voltage on charge that rapidly falls when a load is applied. Equalising may partially help but not if sulphation is severe.

Charging efficiency

With conventional deep-cycle batteries, charging a 100 Ah battery to 100% requires up to 115 Ah. AGMs and gel cells charge well at slightly lower voltage hence their overall efficiency is marginally higher. Lithium batteries maintain an almost constant output voltage (in typical RV use). Their charging efficiency is thus higher still.

Battery capacity & temperature

A lead-acid battery's ability to accept charging current increases with temperature by 3% to 5% for every 5°C above 25°C - and vice versa. The acceptable maximum is 50°C. (Figure 3.11). Check the maker's specifications re this (e.g. AGMs are at risk of being overcharged at temperatures above 40°C.)

High quality chargers and solar regulators have a (usually optional) battery heat sensor that adjusts charging accordingly.

Self-discharge

Whilst today's starter batteries are less affected, unless the engine is run at least once a month or so, they still lose some of their charge, and increasingly so toward the end of their life. If laid up for longer than four to six weeks they are likely to need floating across a small solar module and reliable regulator, or charging once a month or so.

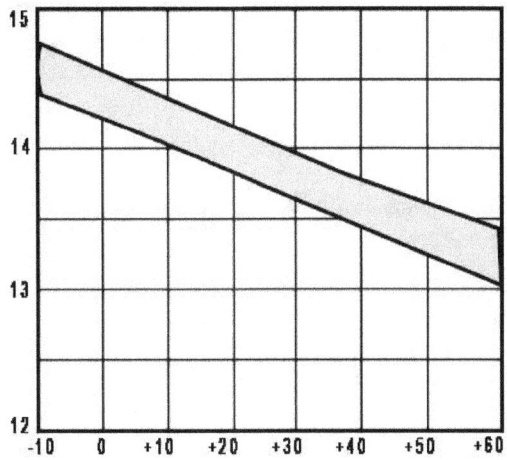

Figure 3.11. Some regulators and chargers reduce voltage as temperature rises. Vertical axis is voltage. The horizontal area is degrees Centigrade. The shadowed area shows acceptable charging limits for lead acid batteries.

Deep-cycle batteries in good condition lose 2% to 3% of their charge per week in winter, and about twice that in summer. Floating across a constant 13.2 volts or so counterbalances self-discharge.

Gel cell and (particularly) AGM batteries have less internal self-discharge. In temperate climates, they may lose only 30% or so a year. Many battery makers suggest that, for AGM batteries not in constant use, to fully charge initially - and then only with, for example 12 volt batteries, if/when they fall below 12.6 volts. Even in hot places, this is usually only after six months. In cool places, it may be a year or more.

While electrically rugged, both AGMs and gel cells can be damaged by long-term float charging because very few chargers can supply and/or control float current at the tiny levels required.

Gel cell batteries are best stored as above but may need charging every six months. LiFePO4 batteries have negligible self-discharge and are best left alone.

Pulsing

This is a still-promoted add-on of (in RV Books' opinion) of dubious value.

The vendors claim that imposing a low level ac current on the battery charging current reduces shedding of active lead-acid battery plate material. It is also done by applying short high voltage pulses.

Now rarely used, pulsing was claimed to enable sulphated batteries to be substantially restored. One vendor (selectively) quoted parts of an Australian CSIRO test: but removed the CSIRO's qualification that 'results were inconclusive'.

Assessing battery condition

The battery industry regards a battery's life-span as that when its capacity is 80% of that when new.

A very rough check can be made by applying 14.4 V across a fully-charged lead acid battery and measuring the charging current flow. If it is higher than 2% of the Ah rating, leave it on charge with 14.4 volts across it for a day or two. If the current is then still above 2% the battery may need replacing. Some suggest 1% to 1.5% but that seems overly conservative.

Lithium

A lithium cell's voltage output is more or less constant 13.1-12.9 volts almost regardless of load in typical cabin and RV use - falling off only marginally as it discharges.

These batteries *must* have a management system that ensures cells are evenly charged. This is not necessarily included. Most require a charger that protects against issues such as reverse polarity and excess charge voltage, and also against discharging below a preset voltage etc.

Management functions are provided in some existing battery chargers. Some vendors, however, attempt to make the battery warranty conditional on also buying an often grossly overpriced charger. This practice (called third-line enforcing) is now illegal in Australia.

Mains voltage battery chargers

Many low-end mains-voltage battery chargers are marketed as 'multi-stage' chargers but most charge at a fixed 14.4 volts or so. They theoretically turn off when the battery reaches that voltage but not all do that reliably, let alone accurately. Many a costly battery has its life shortened by a cheap charger.

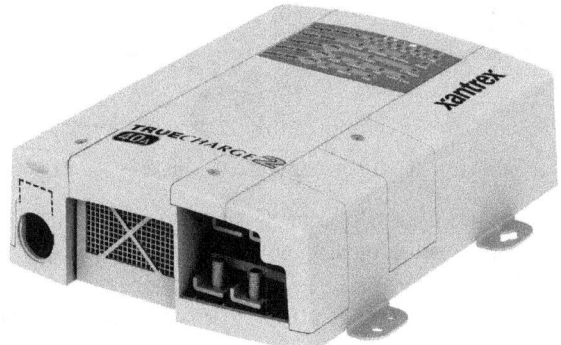

Figure 4.11. Xantrex Truecharge battery charger has settings for conventional, gel, AGM and lead-calcium batteries. It is available with 20, 40 or 60 amp outputs. Pic: Xantrex.

Multi-stage chargers such as that shown in Figure 4.11 are not cheap - but this is not an area for skimping. Such chargers are usually programmable for acid, sealed lead-acid, gel cell and AGM batteries: some also for LiFePO4.

A 20 amp high quality *multi-stage charger may* outperform 40 amp *conventional* chargers once a battery is above 30% to 40% charge.

RV Electrical Converters

Mass-produced RVs worldwide are made assuming that when used, their owners mainly stay overnight in caravan parks where grid voltage power is readily available. Whilst most have a 12 volt dc system, that system consists of an electrical 'converter' - Figure 5.11.

That converter supplies a nominal 12 volts dc (from a grid power ac supply) to *routinely* power the RV's 12 volt lights, water pumps and appliances, i.e. the converter does this *directly* (Figure 5.11). The RV's associated 12 volt battery provides back-up power for that 12 volt system for a few hours in the absence of grid power. (Some US makers refer to it as an 'emergency 'system!)

Where/when grid power is not available (and no other form of charging is available), the associated battery will typically cope with one overnight stay - but rarely two. The battery may then take a day or two to recharge - due to its very low charging voltage.

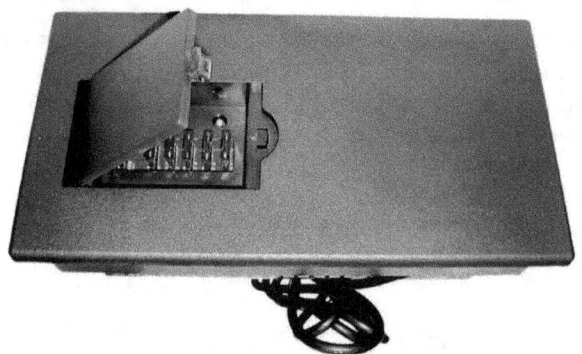

Figure 5.11. Typical 230/12 volt RV converter. Pic: Setec

A typical converter's output is nominally 13.65 volts and drops as load increases. Whilst some converters accept higher voltage alternator input, as seen in Figure 6.11, they have a series diode that prevents the battery discharging through the alternator, but introduces a (typical) 0.6 volt drop.

The resultant typically 13.65 volts is *far too low* for effective charging of lead-acid, gel cells and AGM batteries. It can, however, bring most lithium (LiFePO4) batteries to 80% - 90% charge.

An RV's converter's float charger is thus intended to maintain an already charged battery, and works well for rental fleet RVs to maintain battery charge between rentals.

These converters are cheap but reliable. They are fine for providing 12 volts dc from a 230 volt supply, but few can cope with more than one overnight stay for more than one night.

They are, however, difficult or impossible to modify for free camping use.

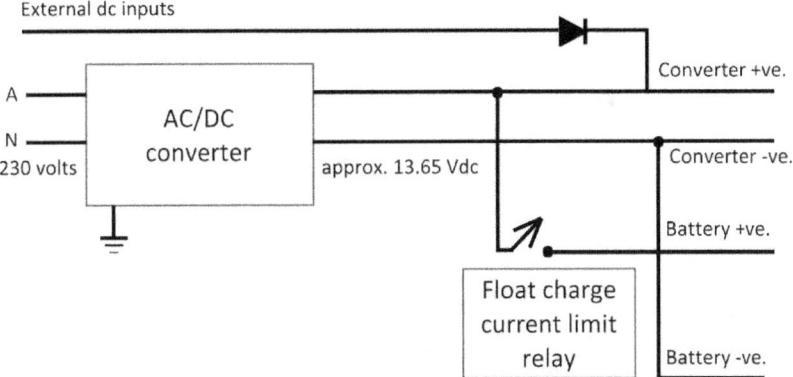

Figure 6.11. A typical 230 volt input converter. Whilst accepting alternator input, the diode introduces voltage drop. Pic: rvbooks.com.au

A partial solution is to install a high output mains charger and increase battery capacity; also required is an dc-dc alternator battery charger (see Chapter 13). This will partially solve the issues in some RVs but, as explained below, not fully in any.

Introduced voltage drop

RV makers know that these converters produce up to 13.65 volts (and that most batteries work at a volt or so lower). Some exploit this to save a few dollars, by using thinner cabling that introduces a volt or so drop. Thus even though batteries are fully charged, energy is lost in that voltage drop.

A costly and only partial solution is to replace the auxiliary battery by a LiFePO4. These charge reasonably well at 13.65 volts and remain between 13.1 volts and 12.9 volts in normal RV usage. That one volt drop will still exist, i.e. the RV's 12 volt accessories will still only 'see' about 12.1 to 11.9 volts - but that's better than one volt less.

Another, but partial, solution is to track down and install a converter that has a regulated 14 volts plus output. They are made in the USA. That inbuilt RV cable voltage drop however remains.

The *only* total solution is to upgrade all charging-related cables and fridge and water pump cabling, and replace that converter by a high quality battery charger. Lighting wiring (if LEDs are used) is usually fine as LEDs are far less voltage conscious than other forms of lighting.

Chapter 12

Battery charging (alternators)

Battery charging technology for vehicles was well developed by 1920. From then, until the 1950s, the vehicle's electrical needs were supplied by a 6.0 or 12.0 volt dc dynamo that generated electricity via coils of heavy wire rotating within a strong magnetic field generated by permanent magnets. The faster the coils revolved, and/or the greater the strength of the magnetic field, the greater the output.

Figure 1.12. Typical pre-2014 vehicle alternator. Pic: Bosch.

The dynamo's main failing was that accessing that output required those coils to be connected to a segmented revolving copper ring, called a 'commutator', from which the current was 'harvested' via carbon brushes. As the commutator and carbon brushes had to conduct the whole of the dynamo's output, heat and ongoing wear limited the power produced. Size and weight were also of concern.

As electrical demand rose, it became essential to develop a more efficient device. The solution was to have the energy-gathering coils stationary, and the magnetic field rotating. This change resulted in the output being alternating current - hence the term 'alternator'.

The alternator's ac output is converted (to dc) by diodes, typically housed within its casing, that allow current to flow only in one direction.

Alternator voltage regulator

As with a dynamo, alternator output relates to how fast it spins - and the strength of its rotating magnetic field.

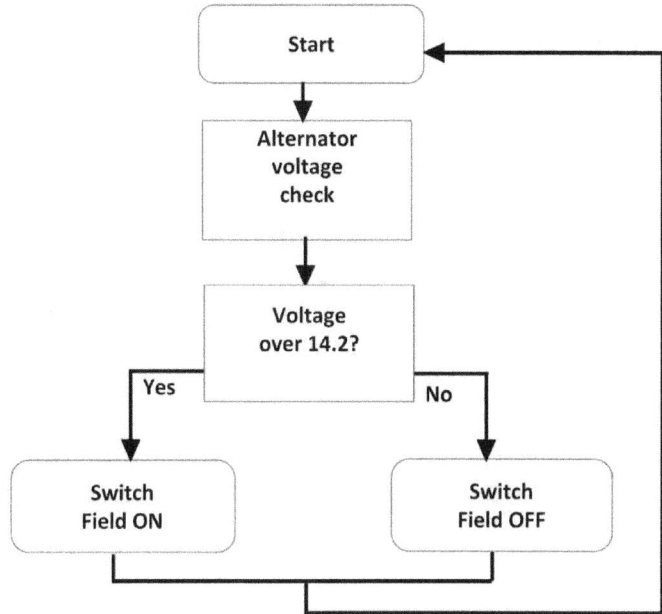

Figure 2.12. How a standard (older) voltage regulator worked. Switching took place at very high speed. Pic: rvbooks.com.au

This field was originally produced by rotating permanent magnets with output controlled by varying the field coil voltage. This did not provide sufficient control so the makers settled for a fully electrical magnetic field controlled by a voltage regulator that originally limited alternator output to a typical 14.2 to 14.4 volts (Figure 2.12).

This output was adequate for recharging a starter battery to the 70% or so of full-charge needed to ensure engine starting, without risk of overcharging in typical driving. It was also adequate for charging auxiliary RV batteries.

In 2000, many vehicles switched to alternators that produced 14.2 volts for some minutes and then dropped to 13.6 volts - too low for rapidly charging auxiliary batteries. Most were thus limited to less than full charge.

With older vehicles, the alternator can often be upgraded to 120 amps but whilst a larger alternator charges batteries *faster,* because its voltage is still controlled by the regulator, only to a similar depth of charge. This is only worthwhile for AGMs, and conventional batteries larger than 150 Ah or so as smaller capacity ones will not accept a higher current charge. Nor will they fully charge without the charging technology described in Chapter 13.

There *were* ways of increasing voltage (by varying the field current), but RV alternators really awaited more effective technology (see Chapters 13 & 14).

Any reliable (pre-2000) alternator is fine for RV auxiliary battery charging providing it is of the right type and voltage, and can be made to fit. Diesel engines usually take a tachometer (revolution counter) signal from the alternator, so it is necessary to retain the original pulley, or pulley diameter.

Lithium batteries can accept massive charge currents (way beyond that of any alternator) but the voltage must be as the battery makers specify.

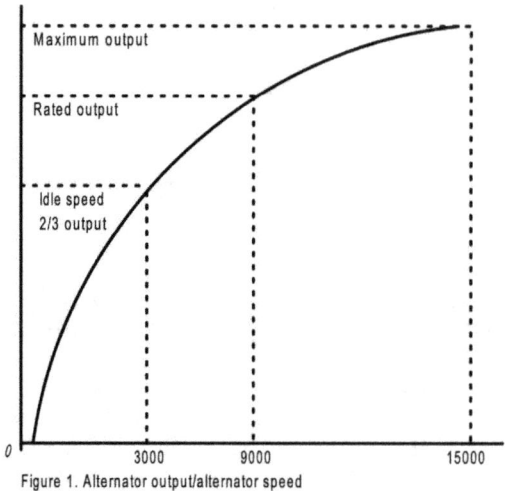

Figure 1. Alternator output/alternator speed

Figure 3.12. Typical output of car and RV alternators. Speeds shown are that of the alternator, not the engine.

Ability to engine restart

While almost all RVs have an auxiliary battery for the RV's 'domestic' needs that battery and the associated system automatically connect across the vehicle alternator, and hence starter battery, for charging. The starter battery thus needs to be protected against discharge while the engine is not running.

Such protection was initially by a high current relay that closed only when the ignition was switched on. When the ignition was switched off the relay opened and isolated the starter battery.

This worked well if the engine was started immediately, but if starting was delayed, a deeply discharged house battery would suck energy out of the starter battery. Then neither (or both) could restart the engine because their paralleled voltage was too low. This became serious with AGM auxiliary batteries because their ability to accept a high charge at lower voltage accentuated the problem - and even more so with lithium batteries. A further issue arose if the load drawn exceeded the alternator's ability to cope, and discharged both batteries.

Voltage-sensed switching

The above issues were resolved by a voltage-sensing relay (Figure 4.12) that delayed paralleling the batteries until the starter battery reached the 13.6 or so voltage required to safely restart the engine. These relays are still available with various current-handling capacities, but vary in quality. Cheap ones have brass contacts that tarnish, causing them eventually to burn out. More costly versions have silver alloy or pure silver contacts.

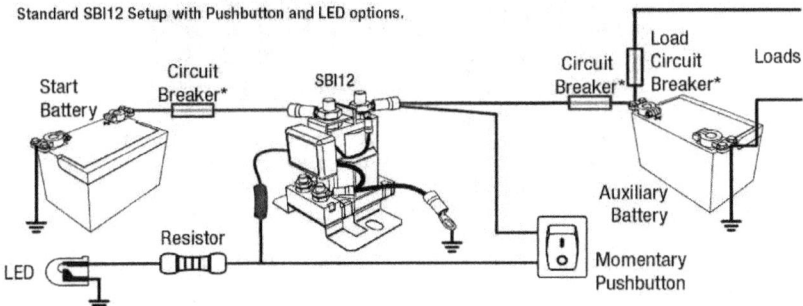

Figure 4.12. How a typical voltage-sensing relay is connected. The (blue) cables, push button, resistor and LED are needed only if there is a need to connect the auxiliary battery manually across the starter battery to aid starting. Drawing: Redarc.

The interVolt relay (Figure 5.12) assists even further. It is adjustable to operate at any realistically required voltage. The relay is claimed to be able to briefly handle 500 amps dc. No such relay, however, is of much value (and some cannot be used at all) with a variable voltage alternator, because that alternator's voltage frequently cycles below voltages of any use for charging.

Emergency starting

For starting a big diesel in cold places, or switching a big electric winch, consider installing a heavy-current (approximately 300 amp) solenoid to temporarily parallel the house and starter batteries.

*Figure 5.12. This interVolt unit can be programmed to switch on and off in 0.1 volt steps from 9 to 38 volts - and with time delays from 1.0 to 255 seconds. It is rated at 150 amps dc continuously, and 500 amps dc for up to 10 seconds.
Pic: interVolt.*

This second solenoid has its contacts wired across the first, and its coil actuated by a dash-mounted switch. Install a warning light that illuminates when the batteries are interconnected.

Always run the engine while winching, and leave it running for about 15 minutes afterwards. If you don't, the temporarily paralleled batteries may discharge too low for engine starting.

Do not parallel solenoids in continual use. If either solenoid's contact resistance increases, the other carries higher current, and then fails.

Regenerative braking

Traditionally, braking a vehicle was done by high-friction brake pads or linings being forced against a drum (and later disc). This caused the vehicle's kinetic energy (i.e. energy inherent in any mass that moves) to be turned into heat. It was simple and relatively effective, although such truck brakes tend to overheat and fade.

Regenerative braking, now used in electric and hybrid vehicles, causes the a vehicles' drive motor to work in reverse: to become a generator when braking is required. This is a particularly effective system in city stop-start driving. Traditional brakes are retained for situations where regenerative braking is not sufficient.

The system is efficient and effective but the brake pedal feels different. It must be depressed further and harder for the friction brakes to be activated.

The need for change

Ongoing emission reductions necessitated computer control of electrical fuel injection, ignition timing and, in some engines, of valve timing. Doing so electronically made it feasible for such interaction to extend to automatic transmissions, traction control systems etc.

Realising the above inevitably required a new approach, various manufacturers worldwide initially researched and developed a technology (dc-dc charging) that electrically isolates such charging from the vehicle's main computer system. The vehicle thus perceives the dc-dc charger (electrically) much as a pair of spotlights or an audio system. It also includes the MPPT technique described in Chapter 11. This was then further developed to cope with the variable voltage type of alternator described in Chapter 15.

The CAN bus

The CAN bus is a digital communications system developed for vehicle use. It enables any manner of CAN bus-compliant systems to be controlled without the previous need to have every individual unit hard-wired via switches and masses of control wiring.

It does this by overlaying digital control sequences from the vehicle's main central computer onto the vehicle's wiring system. It is hugely complex and requires a book at least the size of this one to explain. Its use now extends from engine management systems, ABS and traction control, electric windows and air conditioning, to sirens, flashing lights, seat-belt fastening, hand-brake warning lights etc.

The concept was initially developed by Robert Bosch GmbH in 1983. The first CAN controllers were produced by Intel and Philips. Bosch published the CAN 2.0 specification in 1991. It became mandatory for all cars and light trucks sold in the USA in 1996, in all petrol engined vehicles sold in the European Union since 2001 and in all diesel vehicles sold since 2004.

Many recently-made RVs have this system. It works well and saves cable but can preclude any later DIY additions unless they too are CAN bus compliant, or added as an independent system. For most owners this will be auto-electrician territory.

Chapter 13

dc-dc alternator charging

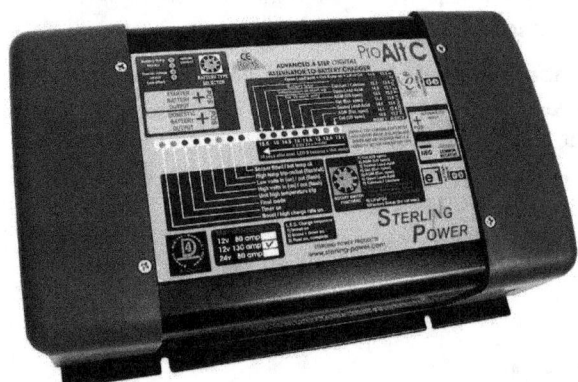

Figure 1.13. Sterling Power's dc-dc alternator charger are available in a wide range of current ratings (both for RVs and boats. The manufacturer has units that charge up to 200 amps plus. Pic: Sterling Power (UK).

Dc-dc charging enables alternator and solar output voltage to be optimised enabling batteries to be charged at the voltage and regime required. The systems typically accept voltage input from 9 to 38 volts and are programmable for virtually any type of battery (Figure 1.13). Some include solar regulation, remote energy monitoring and mains-voltage charging. A few are described by their makers as 'battery management systems' but that's a marketing term. All work in much the same way.

The more complex units cost more initially but are simpler and cheaper to install.

A commonly feared downside is that if one functions fails, then all fail. Furthermore, issues may arise with separate units as they may not be fully compliant.

Most RV electrical issues, however, relate to external connections, and the combined units make most of those connections within that unit. Failure is thus *less* likely. Their installation is also simpler.

Chapter 14

Installing a dc-dc alternator charger

As a dc-dc alternator charger optimises charging voltage, locate it near the associated auxiliary battery. Whilst the unit recognises and compensates for voltage drop across the feed cable that drop nevertheless causes energy to be lost as heat. It is thus still desirable to use adequately heavy cable. This need is emphasised by the charger manufacturers, but some specify only the minimum acceptable size. It is better to use the cable sizes recommended in this book (Chapter 7) particularly if charging large capacity batteries.

A (sometimes optional) remote monitor advises battery state of charge and what is coming in and going out. It can usually be located anywhere conveniently visible, via a USB connecting cable.

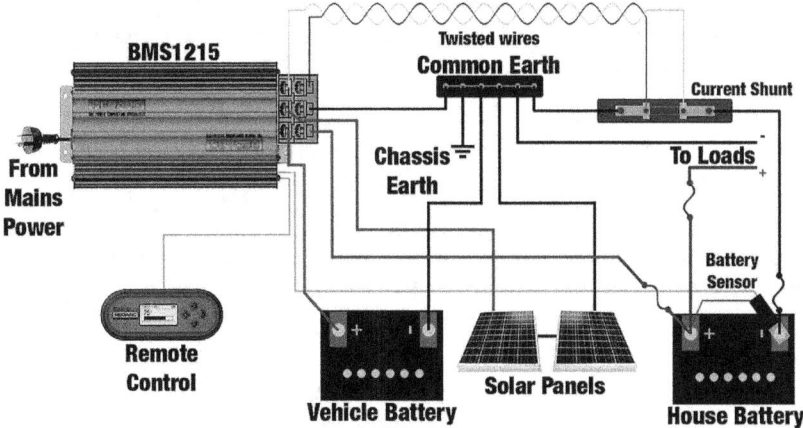

Figure 1.14. Typical battery management system installation. Most interconnections are made within the unit itself. Pic: Redarc.

Chapter 15

Variable voltage alternator charging

The global quest to reduce vehicle exhaust emissions concentrates on any number of ways of reducing the amount of fuel consumed. This includes reducing vehicle weight, ensuring fuel is burnt as completely as possible, reducing all unnecessary loads (including electrical) and recovering any energy that's feasible. These, and other related requirements, are progressively tightened. Vehicle makers must meet emission requirements but are free to choose various ways of doing so. This includes alternator technology that began in 2000, further tightened around 2010 and again in 2014.

There are now three main types of alternator. The vehicle makers choice of which to use is determined by whatever best enables them to achieve the required emissions level.

Fixed voltage: fitted to all vehicles prior to 2000 and many until 2010-2011, these produce a theoretically fixed voltage typically exceeding 14 volts. They never drop below 12.7 volts whilst driving.

Temperature compensating: fitted to many post 2000 vehicles, their voltage varies with engine temperature. It is typically 14 or so volts when cold: it decreases as the engine warms up but does not drop below 13.2 volts whilst driving.

As a generalisation, the voltage sensing relays described in Chapter 12 can be used with alternators that never (whilst driving) fall below 12.7 volts - as with most that are temperature compensating.

Figure 1.15. Typical variable voltage alternator. Pic: source unknown.

Variable voltage: fitted to most post 2014 vehicles, voltage varies from 12.3 volts to 15 volts plus whilst driving. A few fall to zero volts: Figure

1.15 shows a typical such unit.

These variable voltage alternators that fall below 12.6 volts are the ones most likely to cause issues in recently-made vehicles. It is currently necessary to use a dc-dc charger unit that is compatible with this alternator type.

To overcome this issue, the starter battery voltage is used as a reference for what the alternator is doing, optimising RV battery charging accordingly. A dc-dc charger will not operate successfully in such operation. Suitable alternator chargers are complex units made by a relatively small number of makers. An experienced auto electrician can advise.

Identifying alternators

Variable voltage alternators look that shown in Figure 1.15 - they have a very wide or multi-belt pulley. The only certain way to determine your type of alternator is to measure the voltage across the starter battery over a wide range of engine temperatures and driving conditions to see if it ever drops below 12.7 volts. If it does it is odds-on to be a variable voltage unit.

Chapter 16

Inverters

Inverters convert energy derived from (usually) lower voltage direct current to a higher voltage alternating current (in Australia and New Zealand to 230 volts, 50 Hz). Current sine-wave inverters are 95% or so efficient across their mid-range output, and are barely affected by changes in input voltage. Some are inverter/chargers that act as both simultaneously if/when required.

While their function is unchanged, there has been a recent major change in inverter technology that necessitates careful selection to ensure satisfactory operation. There are now two main types.

Transformer-based

Transformer-based inverters use a technology the major component of which (the transformer) enables them to produce two or three more times their continuously rated output for a few seconds, and at least one and a half times more for a few minutes - Figure 1.16. This ability is inherent. It is not an overload. The advantage of this is that they readily cope with some loads, such as induction motors, pumps and almost all heat producing devices, that draw twice or more their running current while starting.

Switch-mode

The later and increasingly common switch-mode inverters are smaller, lighter, more efficient and far cheaper per watt. Few, however, have appreciable long-term overload ability. Many are limited to 80% or even less of their full output if used for more than a few seconds. Largely related to price, some will only supply 50% of 'rated output' after a minute or two, and are almost useless.

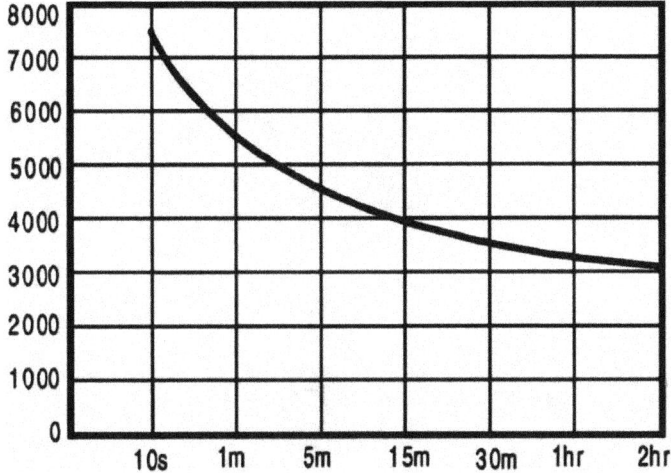
Figure 1.16. Output of a typical transformer-based inverter. Vertical axis is watts.

Where loads have little extra initial draw (such as LEDs, TVs, computers, etc.,) these inverters are very much cheaper than transformer-based units but, if they are to drive anything with an electric motor, etc., you will need one of many times its marketed rating. A switch-mode unit's small size and lower weight may be of benefit but the price will be much the same as a transformer type unit.

Chapter 5 shows the draw of many loads likely to be used in an RV. Total the energy draw of all likely to be used simultaneously, allowing for start-up surges, then check the actual ratings (over time) of switch-mode inverters to find one that copes. Or seek a vendor's advice, plus confirmation that whatever is recommended really will cope.

Modified square-wave inverters

These are cheaper but produce 'dirty ac electric current'. They are fine for many purposes but may damage some electronic equipment (particularly laser printers). An increasing range of equipment will not run from them.

Unless you really do know what you are doing and are buying a unit for a specific purpose, a pure sine-wave transformer (or a switch-mode) inverter is usually a better choice.

Freestanding inverters

Apart from their manner of working, inverters have two main ways in which they may be used. As shown in Figure 2.16, freestanding inverters (typically not exceeding 500 watts or so), usually have 230 volt outlet sockets on their front panel.

*Figure 2.16. Freestanding 180 watts switch-mode Powertech inverter.
Pic: Jaycar Electronics*

Appliances may *only* be plugged directly into those outlets. Freestanding units must not be connected to any fixed wiring such as the RV's 230 volt system (it is both illegal and dangerous). These inverters are readily connected to 12 or 24 volt dc. The safest are 'double insulated' and are usually promoted as such.

Wired in

Most higher power inverters are designed for permanent connection to 230 volt ac wiring. New regulations regarding this (and also generators) are in AS/NZS 3000:2018 and AS/NZS 3001:2008 as Amended in 2012. Installation must be done by a licensed electrician.

Inverter/chargers

An inverter/charger (Figure 3.16) can act in either or both such capacities, e.g. a 1500 watt unit can provide 1000 watts of its 1500 watts for driving a large power tool and 500 watts (about 35 amps at 14.0 volts) for battery charging. Most switch automatically between powering an RV from an inverter, and mains power when available. A few can be paralleled with other similar inverters (and a few with compatible generators) to increase output.

Phantom loads

Many electrical appliances continue to draw power when switched off at the appliance, and always so with supply cord power adaptors. Some pre-2010 appliances may draw up to 15 watts/day if switched off only at the appliance. Older remote controlled TVs, etc., too may continue to draw 10 to 15 watts when switched off only at the remote controller.

Figure 3.16. Outback Power inverter/charger modules such as this 12 volt 2 kW unit can be paralleled to increase output. Pic: Outback Power.

These loads may seem small but a cabin or RV may have several and they draw energy 24/7 - often far more than the device they control. By 2010, regulations in most developed countries restricted standby power of devices sold to one watt, and to half a watt as from 2013 but many cheap imports still have that high pre-2010 draw.

One way to check is via an infra-red thermometer to see if whatever is checked is warmer than objects close to it. There should be no difference.

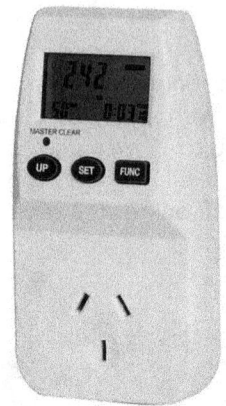

Figure 4.16. An energy meter readily detects phantom loads. Pic: Jaycar.

A better way is via a portable energy monitor (Figure 4.16). Energy meters are readily available from companies such as Jaycar. Unless you are certain there is no draw, turn off-loads at the power socket outlet/s.

Microwave ovens in RVs

Consider your need for a microwave oven when sizing an inverter. A typical 800 watt oven actually draws about 1200 watts - or 1350 watts via an inverter. That oven may cost less than $200 but in small systems adds a thousand dollars or so for the far bigger inverter, more solar modules and extra battery capacity needed to drive it. Consider running that oven only from a generator, or when mains-voltage is available.

Assessing inverter size

As a rough guide, if there is no need to drive a microwave oven, a 750 to 1000 watt transformer type inverter is usually adequate for a medium-size RV. A 3000 watt plus inverter is usually needed for larger RVs, but inverters larger than 1500 watts really need a 24 volt supply. Ideally, put money into energy-efficient appliances rather than a bigger inverter, energy generation and storage or use fewer appliances.

Chapter 17

Installing an inverter

A freestanding inverter (one that has power outlets on its exterior and usually supplied with leads ready to connect to the battery) may be legally installed in an RV by anyone capable of doing so. These inverters must not be connected to an RV's fixed (installed) 230 volt wiring.

For inverters intended for connection to fixed wiring, and all inverter chargers, you may legally do the 12/24 volt wiring yourself, but all work related to the 230 volt connection must be done by a licensed electrician who certifies it accordingly. This applies even if there is no provision for grid or generator 230 volt connection.

A big inverter draws up to 300 amps (as much current as a light starter motor). It needs to be housed as close to the battery as feasible.

To calculate cable size, take the maker's continuous power output for the inverter and add at least 50%. For 12 volt systems, dividing by that wattage by 10 results in amps (and allows for internal battery and inverter losses). For 24 volt systems, divide by 20.

Size the cable, for no more than 0.2 volts drop (or 0.4 volts at 24 volts), by using the formula and conversion chart (Chapter 7). Do the calculations before finalising plans. The cable is likely to be large: you need to make sure there is space for routing it.

Many inverters have cooling fans. Check the noise level before finalising plans as you may need to locate the inverter elsewhere than initially thought.

Connecting an inverter

Battery/inverter cables must be protected against breaking, working loose, or being damaged. It is worth installing a circuit breaker or fusible link (rated at 30 to 40% higher than the maximum continuous current of the inverter) at the battery end of this cable. If necessary have the cables made up by an auto electrician.

Connect the 12/24 volt side by following the inverter maker's installation sequence. If none is given (or available) it is commonly suggested to follow this sequence carefully:

1. Ensure that the circuit breaker is 'OFF' (or the fusible link is not in place).

2. Ensure the inverter is 'OFF'.

3. Connect the negative (black) lead to inverter negative.

4. Connect the other end of that negative lead to battery negative.

5. Connect the positive (red) lead to inverter positive.

6. Connect the other end of the above lead, via the circuit breaker, or fusible link, to battery positive.

7. Re-check steps 1-6 and remedy immediately if incorrect.

8. If all is well, insert fusible link, or click circuit breaker to 'ON'.

9. Switch on and check that all indicator lights, etc., are as per the maker's instructions. Then plug in a 230 volt appliance and check that it works.

Some inverters only switch on above a (sometimes adjustable) minimum load threshold. If so, it may have been set too high. Before altering it, plug in a heavier load. If the inverter now works, reconnect the lightest load you are likely to use and adjust the threshold setting so that it switches on at that level. It is unlikely to be able to detect the tiny draw of an electric clock.

Not all inverters have such adjustment and in these cases there is little one can do except to somehow work around that situation by, for example, using a battery powered clock.

Chapter 18

Generators

Figure 1.18. Honda EU20001 produces 1600 watts (2000 watts peak). At 53 dB at 25% load) at 7 m, it is quieter than most. Pic: Honda.

Whether petrol, diesel or LP gas fuelled, electric generator ratings usually indicate the maximum power in watts that they can produce. With many, that maximum rating can only be used for short periods. They are typically limited to 80% of that for continuous use. They are usually most efficient between 60% and 80% of that continuous use load, but are less efficient if used to power loads below about 40% of that load.

It is far more efficient to run a generator at about 70% load for an hour or two during the day to charge a battery bank via a 230 volt charger in an inverter system - rather than using it to power a TV and a few LEDs at night. This is handy also because generators are banned in national parks, etc.

What appliances draw

Most appliances are rated in watts, both as a measure of energy drawn and of work done. An 800 watt microwave oven produces the heat equivalent of 800 watts but typically draws about 1200 watts. By and large the wattage ratings of simple devices like electric kettles, toasters, etc., are of that when they are fully on.

Electric motors are usually rated in watts and/or horsepower (1 hp is approximately 750 watts). This is a measure of the maximum work they can do. Energy draw relates to work done: it is *always* higher.

Most electric motors draw more current for a second or two whilst starting. A typical 375 watts (0.5 hp) induction motor may draw up to 1300 watts whilst starting under load. Larger motors also draw high starting currents - but not proportionally more. Keep starting currents in mind when assessing the generator size needed (Table 1.18).

Sizing the generator

Rated HP	Rated watts	Start watts	Run watts
0.25	187	750	300
0.33	250	875	370
0.5	375	1300	545
0.75	560	1600	800
1.0	750	2000	1040
1.5	1125	4500	1500

Table 1.18. This data is typical of 230 volt induction motors. Starting current is approximate. It may more than double that shown if starting under load.

A 2 kW generator will run induction motors of less than 500 watts, most power tools, an electric toaster, a big 230 volt RV refrigerator, or a small air conditioner. This about the largest and heaviest load feasible for cabins and small caravans. Larger ones are also generally noisier.

A 3.0 kW generator will run most single phase motors, big power tools, an electric hotplate and most air conditioners. A 5 kW unit will run electric ovens, etc., and many big power tools such as a cut-off wheel or big angle grinder.

To produce the desired frequency, ac generators run at specific speeds related to the required frequency. It is typically 1000, 1500, or 3000 rpm for 50 Hz; and 1200, 1800 and 3600 rpm for 60 Hz.

To check generator frequency, connect an electric clock and check whether it runs fast or slow against an accurate battery-powered time piece. Some generators are speed adjustable.

The universal motors used in power drills, grindstones, vacuum cleaners, etc., are capable of variable speed. They are also more efficient - typically 70 to 80%.

Inbuilt motor-generators

These are big (3 to 7 kW) generators used in large motorhomes and coaches. Some are slow-running, LPG, petrol or diesel-engined, in 'sound-proofed' enclosures. These generators may seem acceptable in urban areas but are not in otherwise quiet campsites. Quieter but costly well-engineered versions are made by Onan and others (Figure 2.18).

Figure 2.18. Onan has LPG, petrol and diesel-engined generators for permanent installation. Pic: Cummins.

Fuel consumption of these generators is usually less than of portable units. Onan claims 1.4 litres/hour at 50% output, and 2.2 litres/hour at full output for its 3600 watt petrol generator. There is an LPG version.

Diesel generators typically produce power for less than A$1 per kWh but cost more than their petrol engine equivalents. Both types are very reliable. They require servicing at similar intervals as the vehicles that carry them.

A generator used without battery backup is hugely inefficient if run for minor loads such as a TV or even a few LEDs. Worse, they must run continuously when driving refrigerators that are typically 'off' half the time. It is far more economic, and quieter, to use that generator for an hour or three a day to power a battery bank/inverter system and, if and only when needed, for extra-heavy loads.

The '12-volt dc output'

Most 230 volt generators also have a 12 volt dc output, typically 8 amps. This output is intended for driving small loads directly. Despite being labelled as a 'battery charger' its typical 13.2 volts will only bring a battery to 50% charge. Despite this, many users run them all day long in (impossible) attempts to fully charge their batteries. Drive a multi-stage battery charger from the 230 volt ac output instead.

Twelve/twenty four volt generators

These generators have a petrol or diesel engine driving an alternator. They output 12 or 24 volts dc only. They are best used to charge a battery bank either directly, or ideally via a high output dc-dc charger, and supplemented by solar. Most are extremely noisy but appeal to people who fish seriously and need a lot of power to drive their big 12/24 volt chest freezers.

Building your own

It is readily possible to build your own dc generator by coupling a high-output 12 or 24 volt car alternator to a suitable engine and ideally feeding the output via a dc-dc alternator charger.

Belt drive and pulleys absorb a kilowatt or so but help dampen the ongoing accelerations inevitable with small engines. These cause damaging high voltage spikes - particularly as the engine splutters to a halt whilst running out of fuel. Adding a filter (made for 230 volts) removes the worst (electrical suppliers will advise that needed.

Vehicle alternators run best at 6000-7000 (alternator) revolutions/minute. To optimise fuel economy, use a pulley size that enables the engine to operate at the peak of its torque curve. This is usually between half and two-thirds of its maximum engine speed.

Chapter 19

Installing a generator

All portable generators (in Australia) must meet the requirements of AS 2790. If permanently installed in an RV all wiring and obligatory safety devices must be done such that it also meet the electrical requirements of AS/NZS 3001:2008, as Amended in 2012 - with particular reference in that Standard to major changes re RCD/CB location etc.

As with freestanding inverters, most portable generators have socket outlets built into them. Appliances may only be plugged into those outlets (via an approved supply cord or multi-outlet board if necessary). it is illegal (and dangerous) to have them connected to any fixed wiring.

Generators must never be operated where their exhaust fumes can be blown or sucked into an open door, window, or air vent as there is a serious risk of carbon monoxide poisoning as a direct result.

While outside the scope of this book, there are now additional and stringent safety requirements for generators used on building and demolition sites, and also on temporary camping grounds.

Stalling on start-up (issues with)

It is not uncommon for a generator to stall while attempting to start a big motor or battery charger if that load is already connected and switched on. In most instances this is because such loads typically draw twice their running current for the first second or two. This is usually solvable by connecting the load only after the generator has started. This may not be possible with big air compressors: some start under load and may need a generator capable of producing up to ten times their running load.

As with induction motors, battery chargers are far from efficient. Worse, their adverse power factor requires a further 30% or more current to be available (albeit not consumed). Both may need a generator of twice the output than might otherwise have been assumed adequate.

A further problem may arise when a generator drives a switch-mode inverter/charger. In some instances the charger will not start at all, may switch off shortly afterwards or not produce full power.

The cause is usually an undesirable generator characteristic that increasingly caused issues following the introduction of a later switch-mode technology predicated on the availability of clean ac. Small generators are directly coupled to single cylinder engines. On their firing strokes they accelerate sharply for a fraction of a second, then decelerate on compression strokes. The rapid speed variations cause voltage spikes, and frequencies above 50 Hz, being superimposed on the intended 50 Hz (cycle/second) output. The resultant 'dirty' output was absorbed by the earlier transformer chargers. It can, however, damage switch-mode chargers, so a protective mechanism within such chargers detects such issues and cuts power. Resultant issues are almost impossible to resolve as each product may work well enough alone. Each vendor usually blames the other - despite it being a mutual problem.

In more recent years, the move to inverter/generators, such as the Honda/Yamaha/Robin/Dometic range reduced the number of such problems but it's still an issue with some $99 specials.

Quietening noisy generators

Quietening a generator is not easy. It is only worth doing to further quieten an already quiet generator. If you house one of these even partly as shown below, you will barely hear it at all. Attempting this is eased by understanding how we perceive noise.

Noise is unwanted or unexpected sound. Quietening it to half its perceived level requires its energy reduced 10 times. Cutting it to one quarter requires its energy reduced 100 times. And so on.

Depending on its formant (i.e. audio spectral make-up), a marginally louder sound from one source may totally mask another sound source that is only marginally quieter. The subsequent sound will seem (and is) different - but may remain perceived as almost or equally loud.

The generator's exhaust is not the only noise source responsible. But, because of that 'masking effect', reducing exhaust noise to almost zero may make little perceptible difference. This is because a lot of the noise from small generators is radiated from their cylinder walls. The exhaust's racket masks the mechanical clatter - and vice versa. *Both* have to be quietened.

There are also differences in how sounds travels. Low frequency sound is almost omni-directional. Higher frequencies are more directional (and more easily dampened). The most effective way to quieten both is to house the generator in a dense limp material that absorbs the sound. By far the best material is sheet lead but it is heavy, costly and dangerous to handle.

Figure 1.19. Sound-proofing foam.

A lightweight and affordable alternative is bitumen-backed acoustic foam. It is available as a vinyl sound barrier combined with a high density sound absorbing flexible barrier mat with an adhesive backing. It is sold by marine suppliers (e.g. Whitworth's Engine Room Deluxe Sound Insulation, catalogue number 80204N - Figure 1.19).

There is a cheaper version of the material but that suggested here is more effective. One or two such sheets lining a plywood box works well. The often suggested egg cartons are almost useless for this. They reduce reverberation within a room but have almost no effect on transmission of any but high frequency sound.

It will also be necessary to add an effective exhaust silencer - many generators have next to none. That from a small car is usually fine. Also required is an adequate air intake into that enclosure. This too needs quietening - via another car exhaust silencer.

As can be gathered from the above, attempting to quieten an existing noisy generator is extremely difficult and virtually pointless. It is far better to buy an already quiet one. (The information is included here as many waste money and time attempting to do so).

Small generator problems

A substantial number of ultra-cheap Asian-built gen sets are still marketed through hardware stores. In non-demanding applications these units may do the job but are liable to damage any electrical equipment that has a switch-mode power supply. They also generate massive electrical spikes if they splutter to a halt by running out of fuel. If you do use one (and they are far from recommended here) install one of the electrical spike filters described above.

These generators have engines that are horrendously noisy and seriously polluting. As there have been quieter and less polluting alternatives since

2000 or so, users will inevitably receive strong and justified complaints from nearby campers.

It is probable that emissions legislation may soon see these generators banned from import.

Chapter 20

Wind power generators

Wind power generators for RV use only *appear* to be a good idea! They have many drawbacks in such usage. Their output is proportional to the square of the propeller's diameter, and the cube of the wind speed. In other words, if the wind speed halves, power output decreases eight times.

Figure 1.20. Aerogen 1.22 metre blade unit produces 120 watts at 37 km/h and 360 watts at 83 km/h (i.e. gale force in some scales). Pic: Aerogen.

These units may spin at wind speeds as low as 11 km/h, but rarely revealed in the brochures few produce useful charging voltage below wind speeds of 20 to 25 km/h. At a more realistic but still uncomfortably strong 30 km/h, the output of a 1.2 metre unit is about 35 watts. Apart from a few coastal areas, however, the average wind speed in Australia is 17 km/h.

Also rarely revealed is that to achieve even the above, they need to be some 30-35 metres above clear ground level.

A 1.5 metre unit *can* produce about 430 watts but one that size (at the necessary 40 km/h plus wind speed) needs seriously strong mounting. In practice, units of less than one metre diameter are pointless for RV use. Larger ones are noisy and dangerous.

Wind run

A wind generator's output varies with wind speed, but it is more useful to check 'wind run' (how *far* that wind travels). The maps (Figure 2.20) show average wind runs for different times of year.

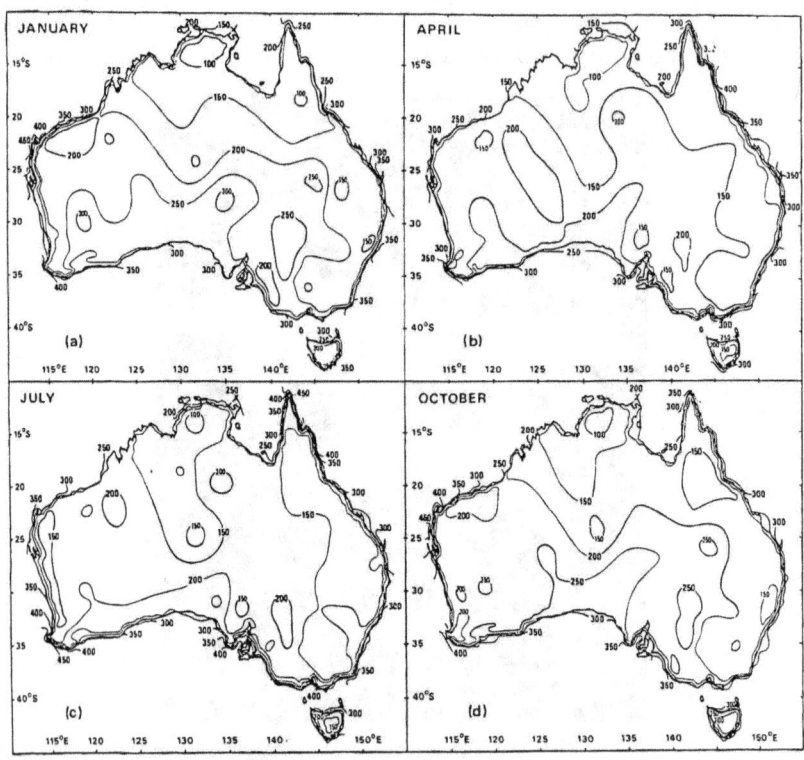

Figure 2.20. Averaged wind-runs (at 2 metres above ground). To convert to metres/second divide by 85. For km/h divide by 25.
Maps: Australian Bureau of Meteorology.

Fishing camps

While small wind generators develop little energy, larger ones can be of value for extended stays on exposed sites, and also coastal sites commonly used by those who fish seriously, and need power 24 hours a day to run their big freezers. In such locations wind generators may be of value.

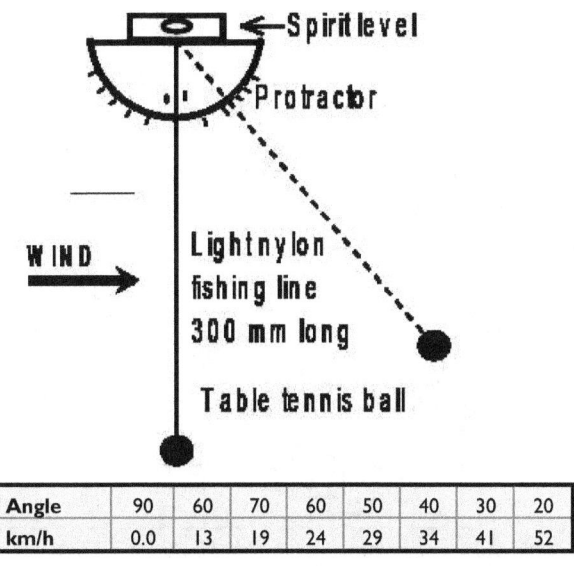

Angle	90	60	70	60	50	40	30	20
km/h	0.0	13	19	24	29	34	41	52

Figure 3.20. How to approximate wind speed. Line length and type of ball must be as shown and the protractor held level.

While small wind generators develop little energy, larger ones can be of value for extended stays on exposed sites, and also coastal sites commonly used by those who fish seriously, and need power 24 hours a day to run their big freezers. In such locations wind generators may be of value.

In many such areas, the wind may often blow strongly and day and night. Figure 3.20 shows how to make an approximation of wind speed.

If a strong enough mast can be arranged, a 2 metre diameter propellor will produce about 750 watts at 40 km/h, and a lot more above that.

Propellor types

All sorts of curious claims are made for propellers. The shape and numbers of the blades affect the speed where useful power begins to be developed but blade diameter is by far the main determinant.

Multi-bladed propellers are claimed to be quieter, (and may well be) but they produce much the same power as if two or three-bladed. Vertical 'egg-beater' type units too develop much the same power as traditional horizontal units - and mounting them high enough to be useful is hard enough in home use, let alone while camping.

Propeller braking

The cube law relationship between wind and propellor's energy is such that strong winds can drive it so hard that centrifugal forces tear off the blades. Automatic safeguarding is essential. This can be done by having a mechanism tilt the propellor out of the wind, or as with electric trains, by applying a heavy electrical load that acts as a brake.

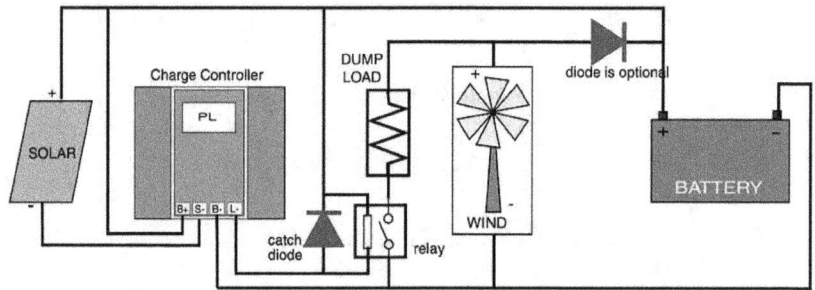

Figure 4.20. A Plasmatronics regulator can be used for shunt operation. Pic: Plasmatronics.

Propellor braking can be done by a shunt operating voltage regulator that controls electrical output by dumping excess energy into a bank of resistors. These, in turn, that convert that energy into heat. Such so-called shunt regulation' is included in some regulators. Shunt operation works well enough as a brake most of the time, but most wind generators and their masts need dismantling for full cyclone protection.

Evaluating & buying a wind generator

There are no official standards for wind generator output. Most makers of the smaller units quote only amps versus wind speed. This is *useless* if the output is for battery charging. It is essential to know the wind speed at which the generator produces meaningful charging voltage (i.e. 14 volts plus). For most wind power generators, that only begins around 25 km/h. To put that into perspective, the average wind speed in Australian coastal areas is 17 km/h.

Much the same applies to claims for maximum power. Qualifications like: 'typical maximum output with voltage limiter inoperative' - is marketing spin for 'one second before it self-destructs'.

As a good general guide, if a camp site has sufficient sustained wind to usefully drive a wind generator, it is too windy for comfortable camping. They also really need to be atop a 30 metre high mast to be truly effective.

Free power (myth of)

The often suggested windmill on a coach roof to provide 'free energy' while driving may seem to be a good idea. It isn't. The alternator-loaded propeller's rotation requires far more fuel to be burned to overcome the drag it causes than you can ever recover in electrical energy.

Installation - do what the makers say

Any wind generator large enough to develop useful power for RV use is potentially dangerous both to erect and to use. Because of this, follow the maker's instructions to the letter. Some warn of the risks but do not provide details about avoiding them.

Chapter 21

Fuel cells

A fuel cell generates electricity from hydrogen. It processes fuel chemically, not thermally. As less energy is wasted as heat, fuel cells can theoretically be very efficient but are currently (2019) limited by needing initially to convert a fossil fuel such as methanol, LPG or diesel, into hydrogen.

If/when hydrogen becomes directly available, as seems probable, they will be about 70% efficient - about three times more so than a petrol-driven generator.

Excepting for a by-product emitted as a small amount of pure water, plus a little CO_2 and heat, fuel cells are almost silent and virtually pollution free.

Brief fuel cell history

The concept of obtaining electricity chemically originated in 1843 but there was little development until fuel cells with platinum catalyst electrodes became used in space and military applications - Figure 1.21.

Figure 1.21. Fuel cell from a Gemini space craft. Pic: UTCPower.

Large scale commercial fuel cells (of a typical 100 kW upward) have been available since the 1980s.

Around 2001, articles began to suggest that fuel cells for boats and RVs would soon be available but most proved to be aimed at raising start-up capital. Many still are.

None became a commercial reality until 2008.

How fuel cells work

A fuel cell's construction is battery-like in that there are cathode and anode plates. Hydrogen is fed to the anode plate and oxygen (from air) is fed to the cathode plate. The hydrogen is split into (positively charged) protons and (negatively charged) electrons via a platinum catalyst.

The protons flow to the cathode via an internal polymer electrolyte membrane.

Electrons flow to the cathode via an external circuit - providing usable electrical energy.

The re-united protons and electrons combine with oxygen at the cathode.

Fuel cell output

Most small fuel cells run on methanol from canisters supplied exclusively by the fuel cell vendors.

It is stated (in Australia) that ultra-clean fuel is essential. Buyers may supply their own but doing so is claimed to invalidate the warranty. It has been argued (if that is so) that imposing such 'Conditions of Sale' is illegal (in Australia).

Typified by the EFOY range, , a typical small fuel typically produces 12 volts at up to 10 amps - (2880 watt hours a day). The units to consume about 0.9 of methanol per kWh generated.

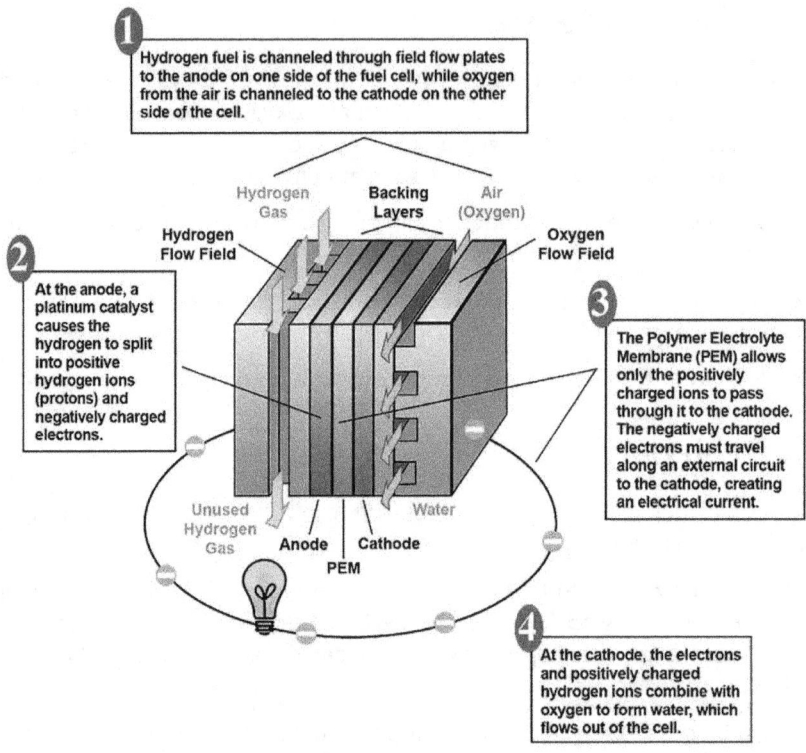

Figure 2.21. How a typical fuel cell works. Pic: Michigan Molecular Institute, USA.

The units were originally handled (in Australia) by Webasto - and cost about A$3200 upward. They were then been sold (and used in their products) by Kimberley Kamper, but are now sold by a number of different companies.

Prices rose from 2012 until 2016 or so but have since fallen substantially. It pays to shop around rather than buying it as an RV 'optional extra'. Googling reveals many sales sources.

LP gas fuel cell

Around 2005, Truma announced it was to produce an LP gas-powered fuel cell developing 12 volts at 20 to 25 amps (about 250 watts - 6 kWh/day). Development took seven years longer than initially announced. Orders began to be accepted in 2012 but, at 10,000 Euros, the otherwise excellent unit failed to find a market. Production ceased in 2014.

Fuel cell efficiency

An issue with both EFOY, and the now defunct Truma, is the need to run on fossil fuels that are then converted (not that efficiently) into hydrogen. This substantially increases their fuel usage - to the extent that, for most purposes, a generator is a far cheaper approach.

Where a fuel cell does score, however, is that (unlike generators that use a lot of fuel even when idling) their fuel consumption is mainly proportional to the load drawn. They are thus reasonably economic for powering only a few LEDs or a small TV if that (and a water pump) are the RV's only electrical loads. At a sound level of a mere 23 dB(A) at seven metres, they are literally all but silent.

As a fuel cells current output is low, but continuous if needed, a battery is required to cope with short-term peaks loads. A low cost 50 Ah starter battery is usually adequate.

Figure 3.21. The EFOY fuel cell has been commercially available since 2009. Pic: EFOY.

Fuel cell future

The commercial failure of the Truma VeGA fuel cell is unfortunate. It was very promising technically. The major market for small fuel cells, however, is not RVs but small settlements in third world countries, and the Truma VeGA was too costly for RVs - let alone for the huge third world markets.

Whilst Truma announced it had ceased all fuel cell development, many other companies are working on various types. One such (the Merlin) is diesel fuelled. There are now one or two new products in the LP gas area but it is too early to comment on them.

Previous editions of this book forecast that fuel cells could well take over from alternators as a source of electrical energy for auxiliary systems in RVs. This may prove *imperative* as emission regulations may eventually preclude using the vehicle alternator for an RV's 'domestic' electrical needs.

Solar however is becoming increasingly efficient and affordable and works well if teamed with a lithium battery and a fuel cell for back-up.

For updates in this area please refer to rvbooks.com.au

Chapter 22

Solar energy

Solar energy is clean, silent and, if the equipment is correctly specified and installed, provides reliable free electricity for years. Where the demand is realistic, it enables RV users to stay indefinitely away from mains power. Chapter 5 shows how to estimate the size. Chapter 23 covers solar installation.

Assessing solar output

On a clear summer day, for an hour or two either side of noon, solar irradiation over most of Australia is 800 to 1000 watts/m². The top commercial modules are typically 20% or so efficient, so allowing also for heat and power mismatch losses, user reality is 110 to 140 watts/m².

To ease sizing the industry averages sun input over a day and expresses that input in Peak Sun Hours' (PSH). The concept is like a rain gauge that 'fills' with sunlight. The 'full' content is 1 PSH. This may require less than an hour in Alice Springs most year-round but half a day in Hobart's winter.

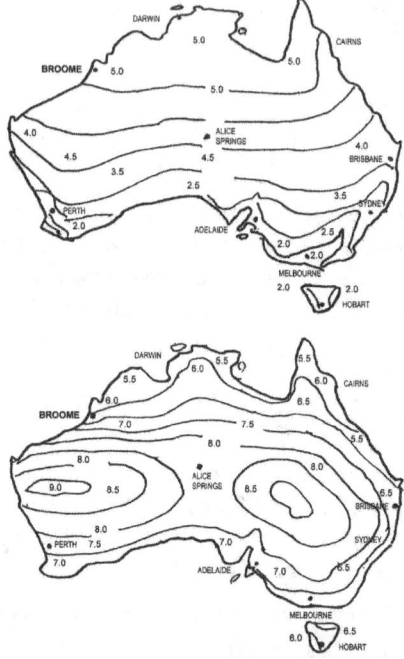

Figure 1.22. Upper map shows PSH for mid-July, the lower, mid-January. Input is roughly linear in-between. Take the PSH and multiply it by 70% of the claimed output of the solar modules (80% if an MPPT regulator) for the most probable daily input. Copyright: rvbooks.com.au

The maps (Figure 1.22) show daily PSH in Australia. A typical mid-summer minimum is about 5 PSH but in mid-winter it only exceeds 4 PSH in the north and by and large varies linearly in-between.

Based on ongoing 10-year running averages, the maps reflect long-term change, not short-term variations. Such data is available for anywhere in the world from NASA.

Solar up north

Solar input in Australia's northern summer is only 15% or so more than in winter - a trap for many who assume it is much higher. This is partly due to water in the atmosphere that acts as a filter.

Night time temperatures are often 25°C to 30°C, so fridges draw 30 to 50% more during the day and 10 to 15% more (and all night too) than in a temperate climate. It is also not unknown for a larger demand for cold beer. Most RVs need about 30% more solar capacity up north than is generally required in summer down south. As a rough guide, unless your batteries are fully charged by midday in southern summer, they will inadequate up north.

Solar module types

Photovoltaic (solar) cells convert light energy directly into electricity. The cells degrade progressively but typically retain 90% of their original output for 20 years.

There are three main types of solar module technology: monocrystalline, polycrystalline and amorphous. Their energy efficiency is related to their type.

Monocrystalline

These single crystal cells are made from pure poly-silicon, doped with boron, and drawn into a single long cylindrical crystal. The crystal is cut into thin wafers, the front surfaces of which are coated with phosphorus. Each cell is thus a single crystal and is usually round, or square with rounded corners.

Figure 2.22. Monocrystalline - originally the preferred choice where space is at a premium. Pic: Victron.

These are the most efficient commercial cells, converting 14%-20% of solar energy to electricity - Figure 2.22.

Polycrystalline

Using a process related to single crystal technology, these cells are cut from a multi-crystal rectangular ingot.

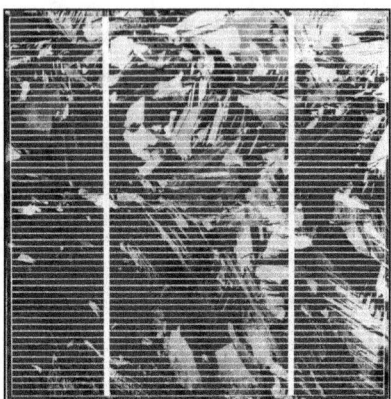

Figure 3.22. Polycrystalline - stunning to look at! The higher quality ones are now as efficient as monocrystalline. Pic: source unknown.

Polycrystalline cells are easier to make and can be assembled closer together. They were orginally only 14% or so efficient and the cheaper versions mostly still are. These are thus larger per watt, and better suited for uses such as stand-alone systems that have ample space.

In recent years however, the more costly versions approach (and a few now equal) the top quality monocrystalline modules.

Amorphous

Amorphous cells increase voltage very slightly as they heat up. Whilst they use much the same energy to make, a drawback for RV use is that these modules are only 10%-14% efficient and hence much larger per watt.

Figure 4.22. Amorphous - thin film and flexible - ideal for curved surfaces. Pic: Powerflex Corp.

Some have a flexible substrate that enables them to follow curved surfaces or be rolled up for storage (Figure 4.22)

Solar module reality

Many people, who know about electrics but not solar industry spin find, to their dismay, that their solar module output is only 70% or so of that they'd calculated.

The solar industry has two sets of scales. One, Standard Operating Conditions (SOC), reflects output achievable only in laboratory testing - not in typical usage. It is nevertheless used industry-wide for marketing.

The other, Nominal Operating Cell Temperature (NOCT), closer reflects user reality.

Since a watt is defined as 1 volt x 1.0 amp, a 120 watt, a 12 volt solar module may be expected to produce 10 amps at 12 volts - yet it typically produces about 7 amps. This is about 85 watts, far less than the 120 watts seemingly claimed.

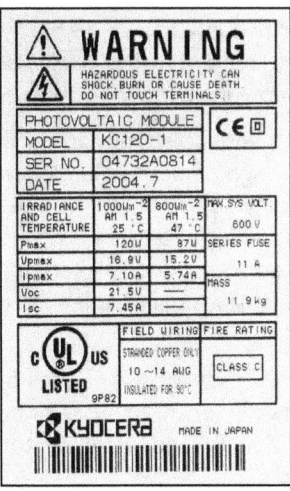

Figure 5.22. The third column shows that the most likely output of this nominally 120 watt solar module is 87 watts at typical cell temperature Pic: solarbooks.com.au.

The explanation is that the industry specifies SOC output by plotting module voltage and current separately. It then defines that output at whatever combination of volts and amps results in the highest number; this is typically around 17 volts. As 17 volts x 7.0 amps is 120 watts, the solar module is marketed accordingly.

The industry justifies this on the grounds that some electrical devices e.g. heaters and electric pumps, accept higher voltage and perform more work; and also that it has long been industry practice.

Standard Operating Conditions include 25°C cell temperature but that is not 25°C *ambient* temperature. It is the temperature of black cells under glass - and at 25°C ambient, that is typically 47°C.

This *is* revealed in the vendors' technical literature, and on a data panel on some modules (Figure 5.22 is an *actual example*). but in industry terms that few buyers are likely to understand. Due to the above, monocrystalline and polycrystalline cells typically lose 4% to 5% of their output voltage for each 1°C above 5°C.

Energy mismatch

A second discrepancy mostly affects older RV and big stand-alone systems and those with older solar regulators. This is a loss due to a battery's requirement for charging each cell at 2.2 to 2.45 volts (13.2 to 14.7 volts for a six-cell 12 volt battery) - yet solar modules develop their maximum power at a typical 17 volts. Via a basic solar regulator, the voltage spectrum between 14.7 volts and 17 volts is not accessible for charging.

The MPPT function described in Chapter 23 partly recovers this voltage gap loss. It began to be included in higher quality solar regulators and chargers around 2005 and is now included in most over $200.

When used in conjunction with solar regulators, the MPPT function automatically accepts a wide range of input voltage: some as high as 600 volts. The MPPT output is typically adjustable for 12, 24 or 48 volt charging, and programmable for various types of battery. This usefully enables solar modules to be series-connected to produce higher voltage at less current but the same wattage, thus enabling smaller solar-to-regulator cables to be used.

The MPPT function typically reduces daily losses by 10% to 17% (the often 30% vendor claim is valid only for 30 minutes or so twice a day - at very low input levels just after sunrise and just prior to sunset).

Fixed & loose solar modules

The drawback with fixed modules is that the vehicle must be in the sun. Solar modules shade the roof from the sun, and the obligatory 50 mm or so air space beneath the modules enables air to circulate, but the RV will still heat up.

Having loose solar modules enables the vehicle to be in the shade and modules efficiently oriented, but precludes charging while sightseeing, shopping, etc. Such modules are also stealable, and damaged by wind gusts unless pegged down.

A further possibility is to mount the modules on the roof of a trailer, which is then left in the sun. Here, it is marginally worth tilting the array to the approximate latitude angle to face squarely into the midday sun but little is lost (except in winter down south) by having them horizontal.

Having distance-located solar modules requires heavy leads, or the solar modules connected to produce 24 or 48 volts, to reduce voltage drop on the interconnecting cable. Some MPPT solar regulators can accept that voltage, and are programmable for charging 12, 24 or 48 volt batteries of various types.

The solar regulator should be connected at the RV end via a heavy-current connector such as the four-pin 35 amp units made for trailer electrical connections, or an Anderson connector.

Dual systems

An ultra-effective approach for conventional and fifth wheel caravans and camper trailers, is to have a separate self-contained (but optionally inter-connectable) system in both tow vehicle and trailer. This also enables a fridge in the tow vehicle to be used while the trailer remains in camp - enabling cold and frozen products to stay that way while shopping.

Figure 6.22. Nissan Patrol and TVan each had is own solar system. Pic: in camp near Mitchell Falls in far north-Western Australia. Pic:rvbooks.com.

A dual system (Figure 6.22) ensures power in the event of the tow vehicle needing repair or servicing. It also enables the living section to be in the shade and the tow vehicle to be in the sun.

Scaling batteries to charging output

Restrictions of space, weight carrying, and/or dollars, limit RV battery capacity. That capacity is limited also to that which can be quickly re-charged. Leaving conventional batteries discharged for long periods damages them - but far less so for gel cells and AGMs.

Nominal battery capacity should not exceed whatever is required to enable the charging capacity to bring the batteries from their typical overnight depth of discharge, to full-charge the next day.

If using solar alone, or primarily, it is advisable to have sufficient solar capacity such that the batteries are fully-charged by about noon (on most spring and autumn days) in coastal areas below a line from Brisbane across to Geraldton.

This is particularly important if you intend to travel in Australia's far north and north-west. If it does not charge as above there will inevitably be power issues up north, even in winter.

Excess charging capacity is never a problem. Over-charging is a function only of excess charging voltage, not charging capacity. There is no risk of over-charging batteries if you have a high quality solar regulator or battery management system that is set up correctly.

Never have excess battery capacity. For RVs, the weight of lead-acid batteries (about 30 kg/100 Ah) limits this anyway. This is less of an issue with lithium batteries as they are not only much lighter, but have lower charging loss.

An RV solar system also requires a regulator that ensures batteries are optimally charged, and a remote energy monitor that provided easy to see information on its (and the battery's) performance. These are covered in Chapters 24 & 27 respectively.

Table 7.22 shows the most probable output for flat-mounted (monocrystalline) solar modules at 30°C ambient temperature)

Modules: 65 watt	1	2	3	4	5	6	8	10
3 sun/hours	120	240	360	480	600	720	960	1200
4 sun/hours	160	320	480	640	800	960	1280	1600
5 sun/hours	200	400	600	800	960	120	1600	2000
6 sun/hours	240	400	720	960	1200	1440	1920	2400

Modules: 80 watt	1	2	3	4	5	6	8	10
3 sun/hours	150	300	450	600	750	900	1200	1500
4 sun/hours	200	400	600	800	1000	1200	1600	2000
5 sun/hours	250	500	750	1000	1250	1500	2000	2500
6 sun/hours	300	600	900	1200	1500	1800	2080	3000

Modules: 100 watt	1	2	3	4	5	6	8	10
3 sun/hours	187.5	375	560	750	935	1125	1500	1875
4 sun/hours	250	500	750	1000	1250	1500	2000	2500
5 sun/hours	310	620	935	1240	1550	1860	2480	3100
6 sun/hours	375	750	1125	1500	1875	2250	3000	3750

Modules: 120 watt	1	2	3	4	5	6	8	10
3 sun/hours	225	450	675	900	1125	1350	1800	2250
4 sun/hours	300	600	900	1200	1500	1800	2400	3000
5 sun/hours	375	750	1125	1500	1875	2250	3000	3750
6 sun/hours	450	900	1350	1800	2250	2700	3600	4500

Table 7.22. Typical output, in watt hours, of flat-mounted (monocrystalline) solar modules at 30°C. The top quality modules now (2019) may exceed this by 5% or so.

Chapter 23

Installing solar modules

Usage	Latitudes <25°	Latitudes >25°
All year round	Angle = your latitude	latitude plus 5-10°
Summer optimised	latitude minus 10°	latitude minus 5°
Winter optimised	latitude plus 5°	latitude plus 15-20°

Table 1.23. The table (valid virtually world-wide) shows tilt angles for solar modules, optimised separately for summer and winter.

Ideally, solar modules should face north (in the southern hemisphere) and be tilted to face directly into the sun. The loss through not doing so varies with latitude and season. For latitudes from 20°C to 25°C or so, flat mounting makes only 10% to 12.5% yearly difference. Anywhere between latitude angle and flat makes next to no difference in the northern parts of Australia.

For RVs, horizontal mounting is simple and effective for all but the south of South Australia, Victoria and Tasmania in the three winter months.

Whilst it is feasible to have adjustable tilting, that may necessitate the vehicle to face in a direction that may not suit other needs. As solar capacity is now so cheap, it is easier to increase overall capacity by 15 to 20% to allow for winter losses and mount the modules flat on the roof. This works better than many expect because indirect solar irradiation often exceeds direct solar irradiation.

Energy calculations rarely result in exact numbers of modules. Unless the discrepancy is truly minor, add an extra module. Assuming space is available, this is simple with paralleled modules and usually possible with modules in series if there is an MPPT solar regulator, as it accepts a range of input voltages. If feasible financially (and it now cheap), have as much solar as possible.

Leave an air space

The output of all except amorphous modules falls off with heat. They need an air gap of about 50 mm above the surface they are mounted on so that heat can escape and cool air flow beneath. Earthing the modules' metal frames to the chassis will reduce dust build up caused by static charges.

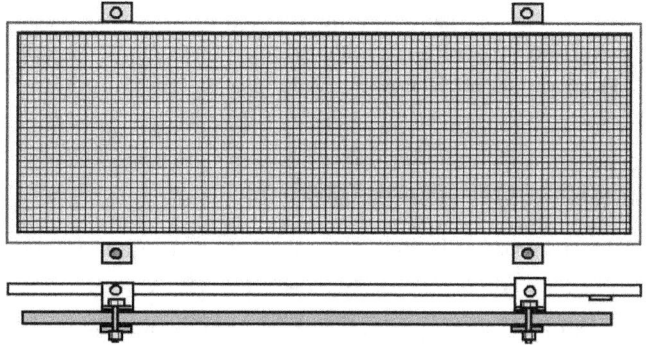

Figure 1.23. Most modules have an alloy frame that can be held down by 50 mm x 50 mm alloy or stainless steel brackets.

Voltage/current

As with batteries, connecting modules in parallel (Figure 2.23 top) increases overall current. They can be of any number and of totally different capacity (a 10 watt module and a 110 watt module are fine) but all must be of the same voltage.

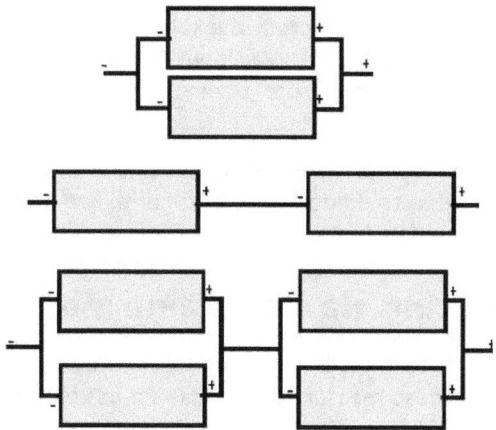

Figure 2.23. Each rectangle represents a 12 volt 60 watt (i.e. 5 amps) solar module.
Top: parallel - 12 V at 10 A = 120 watts.
Centre: series - 24 V at 5 A = 120 watts.
Bottom: series/parallel - 24 V at 10 A = 240 watts.

Connecting modules in series (Figure 2.23 centre) increases overall voltage in proportion to their number and voltage. All must have the same nominal current output. If not, the voltage is still additive but the current is limited to that of the module with the lowest current output.

Banks of paralleled modules may be connected in series (Figure 2.23 centre) to provide whatever current and voltage is required.

As with single series-connected modules, each bank of series-connected modules must have the same current output. No matter how connected, the total energy of the same number and capacity solar modules remains the same.

Measuring module output

Solar modules automatically limit current to their maximum continuous rating. It is not possible to 'overload' them as such.

The methods shown below measure solar modules output directly. Both require you to initially shield the module/s from the sun, disconnect the existing load, uncover to measure and to again cover whilst reconnecting the load. This routine is necessary because making or breaking dc current results in severe arcing that damages screwdrivers, meter probes and (sometimes) people.

Solar measuring via multimeter

For modules suspected of being of less than 10 amps or so, disconnect any existing load and with a multimeter set to its *maximum* dc amps range (usually only 10), measure directly across the solar modules's output terminals. Uncover the module and record solar input (all day if you wish). Cover the module after measuring is complete, then reconnect the original load and uncover the module again.

The above concept may horrify unless you realise the solar modules output is automatically limited to its designed maximum). See below, however, for large or multiple paralleled modules over 10 amps.

Solar measuring via clamp amp meter

Current higher than 10 amps is readily and safely measured by using a clamp amp meter (these measure hundreds of amps but most are accurate to only plus/minus 3%).

Figure 3.23. Clamp meter. Pic: source unknown.

To do so with a clamp meter (Figure 3.23), cover the modules, disconnect existing load and connect a metre or so of heavy cable (about 6 mm² or more) directly across the modules output terminals. Uncover, and measure by opening the clamp meter so that its jaws to accept that cable. Record the result. Cover the module again after measuring is complete, reconnect the load and uncover.

Establishing losses across cabling from the solar array to the actual system can be checked by following that above (including the cover/uncover routine) with the input to that system temporarily disconnected. This will not be exact but close enough if it is done in full sun at the same time both days. By and large the energy monitoring described in Chapter 26 will provide adequate data once the system is up and running.

Solar module voltage measurement

The voltage across a 12 volt solar module is typically 18 to 21 volts when disconnected from the solar regulator, dropping to 15 to 17 or so volts when on load and in full sun.

Solar module care

Solar modules are physically and electrically more rugged than they may seem. Modules rarely fail completely unless torn off their mountings by cyclones, run over by trucks or attacked with rocks.

Faults with solar modules are rare - and usually due to poor or broken connections in the junction box, interconnecting cables and plugs and sockets, or a failed protective diode (see Chapter 23).

Solar modules have a long life - typically producing 90% of their original output after 20 years. They are usually kept clean enough by heavy rain, but in areas like Australia's Kimberley, they can (and are often) partially obscured by blown sand. To clean, use a tiny amount of detergent, a cloth and warm clean water. Rinse, using a teaspoon of detergent in a full bucket of clean water - then let them dry naturally. Never polish them as doing so builds up an electrostatic charge that attracts dust. That trace of detergent in the rinsing water acts as an effective anti-static barrier.

Solar module protection

All solar modules self-limit maximum output. They are not damaged even if the output has a steel bar connected across it. They are usually ultra-reliable but if one cell in a module fails that is the end of that module (most last 30 plus years). None work well in partial shade and all are much the same in this respect on cloudy days. It is rare, however, to experience zero output during daylight hours.

Solar module life

Most solar modules have a life of at 25 years - many last even longer. Output drops slightly with aging. By and it is worth replacing made prior to 2000 as their output per area has considerably increased.

Chapter 24

Solar regulators

A solar regulator's function is to enable solar modules to charge batteries effectively and efficiently. Older regulators monitored battery voltage and cut the feed to the battery when it reached a preset level - but with not always the accuracy or reliability required. Apart from low-end examples (best avoided anyway), most solar regulators are now far more sophisticated and use the multi-stage charging regime outlined in Chapter 11.

There are two main solar regulator types: pulse width modulation (PWM) that chops the solar current input into pulses, varying their width to control output; and multiple power point tracking (MPPT).

Described more fully in Chapter 11, MPPT tracks incoming voltage and current, optimising both to maximise the output for charging and other needs. Charging typically begin with a 'boost (or bulk) mode' at constant current to about 80% charge, drops back to 13.8 to 14.4 volts (absorption) for a few hours, and enters a float mode of 13.4 to 13.8 volts thereafter. Some may include an optional 'equalising' charge (now out of favour with some battery makers and never used for AGMs or lithiums), and/or a brief lower voltage 'warm-up' mode. Apart from the simplest of uses, it is now pointless to use other than MPPT units.

Most upmarket solar regulators include various energy monitoring functions. That built into the best battery monitoring systems (Chapter 26) is typical. If buying a stand-alone solar regulator with an inbuilt or remote monitor), before finalising the buying decision weigh up what is offered against independent energy monitors.

Basic solar regulators have no monitoring. Some makers offer this as an option but if not, a separate energy monitor can readily be added. Energy monitoring truly is needed.

Buying a regulator

The cheapest regulators cost $20 to $100 (depending on current-handling) but lack programmability and monitoring functions. There is then a large price gap. Acceptable units without energy monitoring start around $130 but it really is worth paying $200 upwards.

Figure 1.24. Steca 30 amp solar regulator. Pic: Steca

Solar regulators for wind power generators

Wind power generators must be able to automatically brake the generator's propellor in excess wind. Only a few solar regulators can be programmed for such use (Chapter 24). A specialised 'shunt-regulator' is usually needed: most work by progressively connecting a big resistor bank across the output. (This book recommend against wind generators for RV use.

Programming a regulator

Most regulators are pre-programmed for lead-acid batteries. Time, date, system voltage and battery capacity, etc., are usually easy to set up. Further programming is not hard but the manual may need to be read a few times before it makes sense. If the preset program more or less suits your needs, use it until you become familiar with the regulator's workings and functions.

The battery monitoring systems described in Chapter 26 incorporate the multi-stage MPPT solar regulator functions described here, and also the energy monitoring described later in this book.

Chapter 25

Installing solar regulators

Solar regulators are connected in various ways but generally as shown here. Many have an inbuilt energy monitor and readout, so they tend to be installed at eye level, and thus well away from the batteries they monitor.

To control their rate of charging, solar regulators must know the *exact* battery voltage but any voltage drop along the main feed cables from the solar modules to the battery causes the voltage close to the modules being higher than that close to the batteries (Figure 1.25 - left). The result is that the regulator 'sees' a voltage that is lower than that of the solar array, but higher than that of the battery, and cuts back the charging voltage.

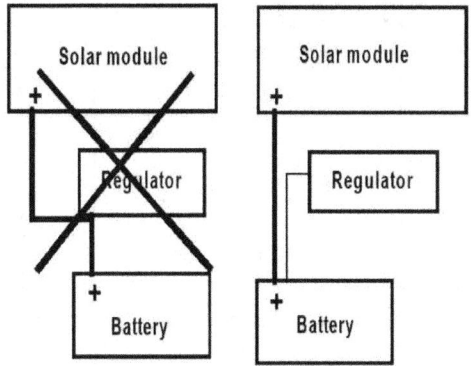

Figure 1.25. If wired as left, the regulator's voltage reference is part way between solar output, and thus higher than battery voltage. Some makers use a dedicated monitoring cable - shown below right. Pic: rvbooks.com.au

Higher quality regulators have a light-gauge dedicated battery voltage cable (Figure 1.25- right) that overcomes this problem, so their location is not critical.

With some regulators, the opposite happens. If that sensing cable is omitted: the batteries overcharge.

A few (even professional) installers appear not to understand the need for this monitoring cable. It seems to duplicate (but doesn't) the main regulator to battery cables. They ignore the instructions and devise ways of leaving it out - typically by parallel connecting the battery and monitoring cable connections at the regulator (Figure 1.25 left).

Some low price regulators have no monitoring cables. Using a cheap regulator is never a good idea but if there is no choice, locate it as close as possible to the battery enclosure (but not inside it) via extra heavy cable to minimise otherwise inevitable sub-optimal charging.

Monitoring issues

Locating a solar regulator that has an inbuilt monitor usually requires the negative or positive cable to all loads to be connected to a 'Load' terminal on that regulator. This is simplified by running a heavy cable from that terminal to a few terminal posts or junction boxes, and connecting the loads to such points.

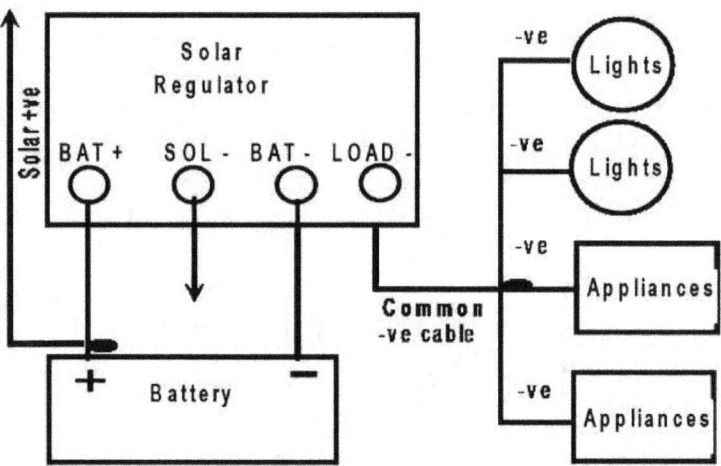

Figure 2.25. Where multiple loads must connect to a regulator's terminal, do it as shown here. Pic: rvbooks.com.au

A major limitation with some such regulators is that the system may be able to handle a hundred or more amps (e.g. to power a microwave oven) yet the solar regulator output connection is limited to a fraction of that. Furthermore, many lack the full range of information required, and/or are difficult to interpret.

A far better solution is to power all loads directly from the battery via a current shunt that handles heavy current and feeds a proportional signal to an independent energy monitor.

A solution for much of the above is described in Chapter 27.

Chapter 26

Energy monitoring

Knowing battery and charging status is necessary for all but basic systems but, due to the ultra-slow reactions between electrolyte and lead plates, attempting that via voltage measurement alone is not practicable. For a big deep-cycle unit, doing so takes 24 to 48 hours, yet is still only approximate. Because of this inevitable time lag, voltage measurement alone cannot tell you what you want and need to know, i.e. the state of charge at any time. Nor is it feasible with lithium batteries as they maintain almost constant voltage across 90% of their usable working range.

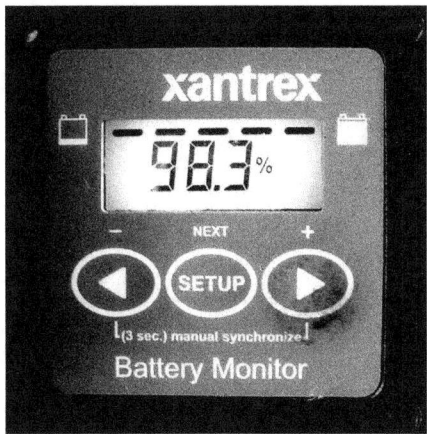

Figure 1.26. Simple to read and very effective. The Xantrex Energy Monitor (also marketed under other brand names). Pic: Xantrex.

Furthermore, and particularly where solar is involved, it is necessary to know how much such energy is coming in, and also how much energy is going out.

Energy monitoring solves this problem. It works much as you track your bank balance. It records energy fed into and drawn out of the battery and deducts the losses. The result is what the battery should contain right now.

This facility is built into upmarket solar regulators and battery management systems. Stand-alone monitors also work this way.

For all such monitors, the 'charge remaining' measurement tends to drift over time but the units usually reset every time the battery reaches and maintains full-charge. Measurement is typically accurate within 5% to 10%. This is adequate for most needs.

Most energy monitors show a readily understood percentage of charge as well as any, or all of, remaining amp hours or watt hours plus incoming and outgoing amps, total daily amp hours, real-time current and voltage, and often the mode of charging (e.g. boost, absorb, float), etc.

Many energy monitors store data over 31 days and enable it to be stored on a laptop computer.

Even if of no personal interest, this data assists any required fault finding.

Chapter 27

Installing energy monitors

An energy monitor accepts data from all related sources and displays all the information reasonably needed. It well worth having one as most solar installations already have inbuilt energy monitoring but display it in terms that virtually require a physics degree to be of meaning. Others may accept and show only the solar input, not that from the alternator or a generator.

Unlike such monitoring, most stand-alone energy monitors display the needed information in a manner that almost anyone can understand - but may also show it in an optional more technical manner for service people to read if necessary.

Figure 1.27. Remote readout for the Redarc BMS (Battery Management System) can be located as shown here. Pic: Redarc.

The need for such monitoring may not be obvious at first - but will become so over time.

Current shunts

An energy monitor requires that all energy generated and used is measured electrically. As the energy used by some appliances (such as microwave ovens is truly substantial it is not feasible to have the massive cables otherwise required to and from the energy monitor. Doing so would also involve energy losses. Instead all current flowing in and out passes through a so-called current shunt (Figure 2.27).

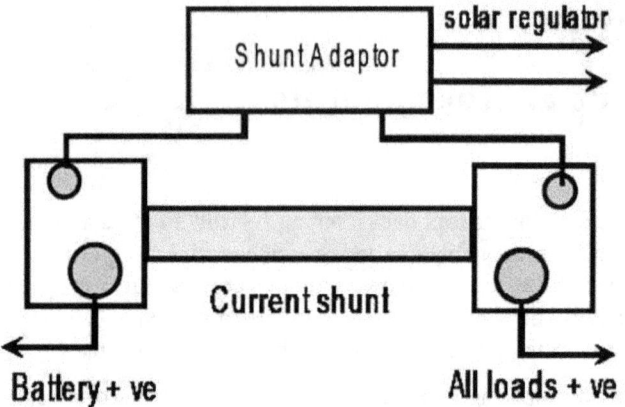

Figure 2.27. A current shunt connects in series with a main battery cable (positive or negative). All loads and almost all energy inputs are connected to the non-battery side of the shunt. (See main text re 'exceptional connection'.) Pic: rvbooks.com.au

A current shunt is simply a few centimetres of metal of known resistance inserted in one or other auxiliary main battery leads close to the battery. One side of the shunt is connected to the battery, and all loads and inputs to its *non*-battery side. All current flowing into and out of the battery (hence through the shunt) produces a tiny voltage proportional to that current. This voltage signal is fed to the energy monitor and shown in amps.

Current shunts for RV use range from 0 to 50 amps to 0 to 200 amps or more. You need one that can comfortably handle the maximum load or charge current and is compatible with the monitoring device.

The shunt need not carry the starter motor draw. Whilst that draw is 500 amps or more, it is only for a few seconds. The typical energy is less than five watts - to low to matter.

The supplier should be able to advise the shunt size and specifications (if not a 200 amp shunt will be fine). The shunt may usually be inserted in either the positive or negative cable but some solar regulator makers specify which. The current monitoring signal is taken from the shunt by a pair of light twisted leads or via a USB cable.

If the shunt is used to extend the measuring range of an existing regulator with an inbuilt monitor, a small shunt adaptor may be needed between the shunt and the regulator. To be sure of compatibility, do buy both shunt and adaptor from the solar regulator vendor.

Figure 3.27. Typical current shunt. Pic: rvbooks.com.au

The maximum permissible distance between shunt and monitor is usually about 10 metres. This will be specified in the maker's instructions. Keep the signal cable away from all other cables.

Exceptional connection

If using a shunt to extend the range of a solar regulator readout, the solar feed output from that regulator must go directly to the battery (or its associated power post). It must bypass that shunt - or that solar output will otherwise be recorded and shown as twice that which it actually is. This has, on several occasions, resulted in hilarious claims on forums for 'how to double your solar input'.

Shunt and energy monitor installation requires considerable electrical knowledge and experience. Do not attempt this yourself unless you have that background. The system may *appear* to be working but is likely to be producing almost random data.

Chapter 28

Lighting

The amount of light needed depends in part on an RV's interior colours. By reflecting light, white or lighter colours saves energy - and increase the perception of space. Dark timber, and any beige finish sucks light like a spectral vacuum cleaner.

Incandescent lights

Lights vary considerably in efficiency. They are literally resistors that run hot enough to emit light (about 95% heat and 5% light). It finally made no sense to use them and the sale of 230 volt versions is now banned in many countries.

Halogen

Halogen globes (Figure 1.28) are about twice as efficient as incandescent globes (and half that of compact and regular fluorescent equivalents). The production of these globes has ceased and with rare exceptions, such as oven lights (high temperature) and particular theatrical lighting all banned from sale in September 2020. Huge numbers are (sadly) still in use - and this topic is covered here for that reason.

Figure 1.28. A breakthrough in their era, halogen globes like this are now being replaced by LEDs.

In their more common 12 and 24 volts forms, halogen globes were produced in 10, 20, 35 and 50 watt sizes. There are two main types: the MR 11 halogen globe has two thin pins 4.0 mm apart. The higher wattage MR 16 has two slightly thicker pins 5.3 mm apart. All now have LED plug-in replacements.

Fluorescent lights

Fluorescent lights use half the energy of halogen lights but need 230 volts ac to drive them. If driven from 12 volts the necessary inverter detracts from efficiency but they still use less energy than halogen globes. Warm white tubes and globes have a pleasant glow, similar to incandescents. They flicker slightly but only a few people routinely detect it.

Figure 2.28. Philips 23 watt fluorescent. Pic: Philips.

Compact fluorescents have inbuilt frequency raising that increases efficiency and eliminates visible flicker. They range from five watts upwards. A 12 watt version produces much the same light as a 50 watt incandescent. They have a wide spread of light so while location is not critical, light may be directed where not necessarily needed.

Some '12 volt' fluorescents are 230 volt lights with tiny inbuilt inverters. Many have a harsh white light, and the cheaper ones generate a lot of radio and TV interference. Higher quality units are available. Some have warm white tubes.

Here too there are now more efficient LED tubes that are direct replacements.

LEDs

LEDs are becoming increasingly efficient. The current (2019) top Cree units produce over ten times the light of an incandescent globe.

Most early LEDs were ultra-efficient and effective as narrow beam torches, spotlights, cooking over campfires, reading, etc., but less so for general illumination. Now, however, they are available and suitable for virtually all uses.

Figure 3.28. There is now a huge range of LEDs such as this 5 watt MR 16 base).

The 12 volt MR 11 and MR 16 LEDs (that also replace halogen globes) are relatively heavy. They are less suitable for RV use, as their thin pins rely only on friction to keep them in place. They tend to drop out of their sockets on any but totally smooth roads. There are, however, fittings that secure such LEDs in place.

If your cabin or RV has grid voltage via an inverter, the ideal is 230 volt LEDs that have GU 10 bases (larger two-diameter, with locking pins - Figure 4.28).

Figure 4.28. The GU10 base will hold a GU10 light globe in place even on severely corrugated dirt roads. Pic: Jaycar.

Elegant LED light fittings are available but at a high cost. Many people however have made affordable and effective general LEDs lights from existing light fittings.

Strips of multiple tiny LEDs are readily available. Some run from 12 volts and draw only a watt or two. Check light colour before buying as many are virtually pure white that some find visually uncomfortable. Most LEDs are available in warm white, daylight white and a few part-way between.

LED light output

Prior to LEDs we somehow 'knew' how much light to expect from (say) a 50 watt globe. All were much the same regardless of brand. LEDs are not like that, they vary hugely in efficiency. Because of this a high-quality 5 watt LED is likely to produce a lot more light than that of a 10 watt eBay cheapie - and last years longer. It is best to buy LEDs from lighting shops - or stay with the top brand names.

Figure 5.28. Direct LED replacement for 230 volt incandescent screw base globes.

The related unit of light output is now the 'lumen' and most LED makers disclose the lumens per watt. Light fitting suppliers are well aware of the confusion regarding this and have displays enabling comparisons to be made.

Installation

LEDs are so much more efficient and the better ones so reliable that there is no point in installing anything else. They use only a fraction of the energy for useful light than any other lighting form.

Such ultra-low current consumption also makes them ideal for older RVs. These almost invariably had inadequate lighting cable that will now be fine for LED's far lower energy draw. For new wiring, 1.5 mm² cable is strong enough not to be accidently broken, and suitable in every other respect.

Compact fluorescent lighting requires 230 volts and may only be installed by a licensed electrician. See Chapter 8 regarding the need to physically separate 12/24 volt and 230 volt wiring.

Chapter 29

Water

There is a valid case for using some mains-voltage equipment in recreational vehicles, but not as yet for water pumps. Mains-operated water pumps are designed for flow and pressures that far exceed RV needs. Further, unless the correct mains-voltage pump is used for the RV's specific needs, overall efficiency may be reduced to as little as 10%. The pumps described below are all designed to run from 12 or 24 volts dc, are readily obtainable and made for marine and RV use.

Make	Volts	Amps	Flow l/m	Pressure kPa (psi)
Flojet 4405	12/24	3.9 (12 V)	11	137 (20)
Flojet 4325	12/24	6.3 (12 V)	14	137 (20)
Jabsco 44010	12	4.0	9.5	137 (20)
Jabsco 36800	12	6.0	12.5	133 (20)
Whale EF0612	12	3.9	7.0	212 (32)
Whale EF1012	12	4.2	10	212 (32)
Shurflo 2088	12	7.0	10.6	315 (45)
Shurflo 2093	12	5.5	7.5	210 (31)
Shurflo 4008	12	7.5	11.4	385 (55)

Table 1.29. Typical 12/24 volt pumps. The Shurflo 2093 is much quieter than most. The Shurflo 4008 is a constant pressure unit.

Makers of pumps that operate from 12/24 volts normally quote their products' energy draw in amps, not watts, but if they do specify watts (and, as with 230 volts ac pumps) that wattage is likely to be a measure of the work they do when pumping - often designated as P_2. The energy draw in watts is shown as P_1.

The simplest pumps are cylindrical and have centrifugal impellers. They are not self-priming so must be located within the tank, or otherwise below it. Their seals are damaged if accidently run whilst dry. Diaphragm pumps are usually more efficient, are self-priming, and are not damaged if run dry. Their small working parts, however, are readily blocked by debris, so they need an inlet-side filter located where it is removable for cleaning.

Few extra-low voltage pumps are as rugged as their mains-voltage counterparts but most run for two or three years before needing a new diaphragm and/or valves. It is thus worth carrying a repair kit, or a complete spare pump. These pumps don't like being unused for long periods. Give them a workout from time to time if the vehicle is not in regular use.

Controlling the flow

Most RV water pumps have pressure operated switches. When a tap is turned on, the pressure fall is detected by the switch - that starts the pump. When the tap is turned off, the pump continues to build up pressure until the switch drops out, and cuts the power to the pump.

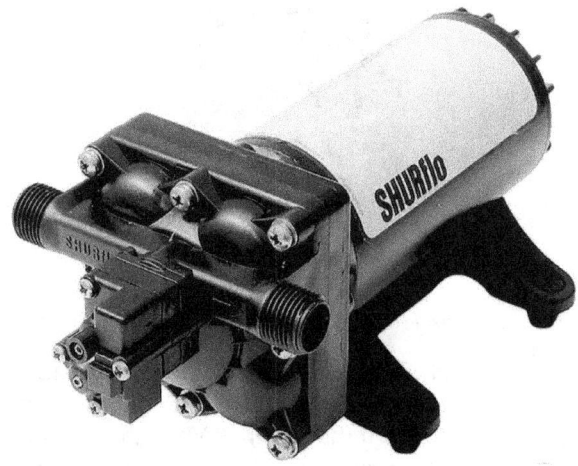

Figure 2.29. Some pumps, like this Shurflo 4008, have an internal loop around which water is pumped when a tap is turned on. That required is diverted from the loop. This ensures constant pressure but uses more energy. Pic: Shurflo.

The switch typically cuts in at about 140 kPa (20 psi), and out again at around 245 kPa (35 psi).

This pressure differential restrains the pump from continuously 'hunting' but it still operates every second or so when drawing a lot of water.

There is however a growing move to both variable speed pumps and those that maintain constant pressure by recirculating that unneeded within the actual pump (Figure 1.29).

Accumulator tanks (pressure vessels)

An accumulator tank is a pressure vessel that reduces pump cycling and smooths water flow. The tank contains a balloon inflated to about 210 kPa (20 psi) - i.e. about 35 kPa (5 psi) below the cut-in pressure of the pump (Figure 2.29). It can be installed anywhere that is convenient between the pump and the taps.

Figure 2.29. Inside a water pressure accumulator. Pic Flexcon.

With the balloon inflated, water pumped into the accumulator tank compresses it to the maximum system pressure. The balloon exerts pressure on the water, forcing it through the system. When the balloon's pressure falls below the pump switch's cut-in point, the cycle is repeated.

With an accumulator tank, a pump cycles for longer but far less often. Energy consumption is hugely reduced because 50% of the volume of the tank can be drawn without triggering the pump (as electric motors draw more current while starting). Pump wear is reduced, and the pump is less likely to chirp annoyingly at night.

For systems with a water heater, these tanks virtually eliminate the otherwise constant and sudden changes in water temperature.

The tanks are made in various sizes: from 1 litre to well over 500 litres. The larger they are the better they work - larger ones are highly recommended for motorhomes and coach conversions that have sufficient space. For cabins, having one of 100 or so litres is well worth considering.

A pump's start-up current is about twice its running current so specify cable accordingly. Connect via a switch to enable the pump to be switched off when you leave the cabin or RV for more than a day or so.

Pressure switches are built into the pump housing or attached to the pump. Others are plumbed into the system at any convenient place between the pump and the first junction that feeds the taps.

Replacing that pressure switch may cost only marginally less than a new pump. It is usually possible, however, to bypass the inbuilt pressure switch by installing an external (and far cheaper) version and rewiring the (usually external) pump pressure switch lead via that new regulator.

To quieten the pump, allow for both the input and the output hoses to form loose loops and only loosely tied such that they have freedom of movement. This considerably reduces impulse noise. Likewise secure the pump to a truly solid base that will not transmit noise, or use soft rubber grommets.

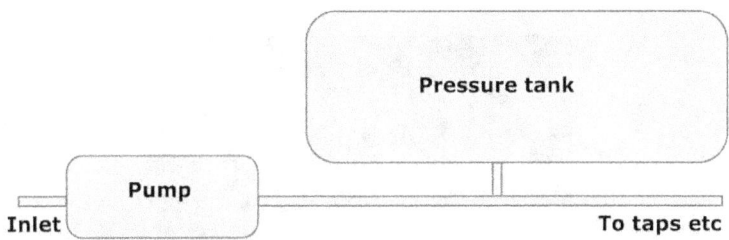

Figure 3.29. Installing a pressure accumulator. Pic: rvbooks.com.au

There are no restrictions on how close the pressure vessel (accumulator tank) may be to the tank or taps, etc. It is typically teed in as shown in Figure 3.29. Some have a separate inlet and outlet.

To set air pressure, check the cut-in and cut-out pressures at which the pump's regulator operates. Then with the balloon empty, inflate it to about 14 kPa (2 psi) below the pumps cut-in pressure and turn on the pump. The pump will run until the pressure reaches the regulator's upper limit. Compressed air pressure then provides the water pressure until the former drops to the regulator's cut-in point and the pump once again turns on until full pressure is restored.

Filters

Coarse water filters must be on the inlet side of the pump, fine filters (a 10 micron and a 1 micron are fine) must be on the outlet side of the pump.

Chapter 30

Electric toilets

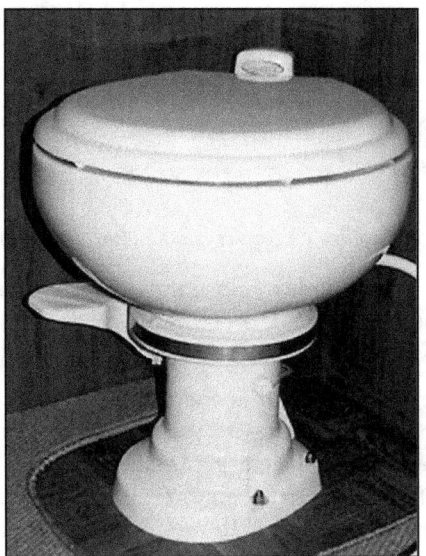

Figure 1.30. Dometic 1600 vacuum loo.

Vacuum operated toilets, made by Dometic and others are now recommended by many major RV builders and users. They maintain a vacuum below the water within the bowl. When the toilet is flushed (by a foot pedal or push button) the vacuum is released. The waste is sucked rapidly down the vacuum pipe and into the waste tank, breaking up waste matter along the way. After the flush lever is released, the vacuum pump continues to run until the vacuum level is restored. Emptying the holding tank is done via a hose from the outlet into the dump point, and switching on an inbuilt discharge pump. The unit shown (Figure 1.30) is the Dometic 1600 series but there are various models. Most draw about 5 to 7 amps at 12 volts while flushing.

Macerator systems

Electric toilets in older camper vans and motorhomes can cause many and varied problems if they have been modified from hand pumping to electrical operation by installing a macerator.

A macerator is a sort of in-line blender that breaks down excreta to about 3 mm diameter so that it may be pumped into the holding tank. Some are combined with a pump that pressurises water for flushing. Others break the stuff up as it exits the holding tank.

Twelve-volt macerators draw 25 to 35 amps so a poor connection stops them in mid-maceration - as will a battery unable to deliver that amount of current, and/or excess voltage drop across the cable. Macerators are not intended for continuous running. Their duty cycle is two to 10 minutes.

Common causes of problems include a faulty timer, or not realising there must be one, by lack of use, chewed-up rubber impellers, odd things flushed down it etc. Some use the possibly preferable Greek/Turkish method of a container for 'used' loo paper. (Thankfully this book is mainly about electrics.)

The electrical part of installing the toilets described above is straightforward. Leave sufficient cable to enable the whole assembly to be withdrawn for subsequent and inevitable maintenance (preferably by someone else).

It is essential to use at least 8 mm^2 cabling for a macerator pump as their current draw is very high. Many issues with these pumps is caused by cable that is inadequate.

The plumbing is best left to the supplier's recommendation.

Chapter 31

Refrigerators

Refrigerators pump heat from where it is not wanted - to where it does not matter. All work like this but there are major differences in how they do it. There are also substantial differences between actual and intended performance in varying ambient temperatures.

Absorption (three way)

Absorption cycle cooling involves heating a (now) water-based solution at high pressure until it vapourises. The vapour is then cooled until liquid and flows into an evaporator in the refrigerator's interior, absorbing heat as it does so. The absorption cycle requires external heat, nowadays via an LP gas flame or an electric element.

Figure 1.31. The Chescold RC1180 has a 30 litre fridge plus 16 litre freezer. It converts to a 50 litre fridge by removing the divider. Pic: Dometic.

It was invented by French scientist Ferdinand Carré in 1858. The first commercial units were made by AB Arctic, a company bought by Electrolux in 1925 that sold the product as Dometic.

Often known as 'three-way' fridges, the RV variants usually runs on 12 volts dc while driving and for short roadside stops, and on mains electricity or LP gas at all other times.

The Chescold unit (Figure 1.31) will run for about 25 days from a 9 kg LPG cylinder, and longer if used as a fridge only. At 10 amps draw, it uses too much energy to run from 12 volt RV solar. The more typical larger three-way door opening units draw up to 25 amps.

Historically, caravan owners have been more likely to experience problems with three-way fridges. This is because very heavy cable is required to avoid excess voltage drop from the tow vehicle alternator to the caravan battery, and far from all caravans have that. As a direct result many three-way fridges performed badly and thereby gained an undeserved poor reputation. To work to specifications, three-way units *must* be installed correctly. See Chapter 32.

Three-way fridges slash electricity usage, and save on solar modules, battery storage, and/or generator fuel. A three-way unit running on gas is a good proposition where there is no space for solar modules, and/or where finances are tight; particularly for extended stays away from mains power. Gas costs partially balance out by savings on battery replacement.

Three-way fridges, in particular, lack ability to freeze a large mass of fresh fish, etc. in reasonable time. They are not recommended for those intending serious fishing.

Compression type

These go back to Persia's ibn Sinai (aka Avicenna) in the 11th century, shown as practicable by William Cullen in 1748, and then by Oliver Evans in 1805. They were first mass-produced, by Frigidaire and Kelvinator around 1918. They compress a gas refrigerant to a liquid, generating heat in the process. The heated liquid is then cooled thus releasing the pressure and allowing the liquid to vapourise. The vapour absorbs heat from the refrigerator and that heat is released to atmosphere via cooling fins.

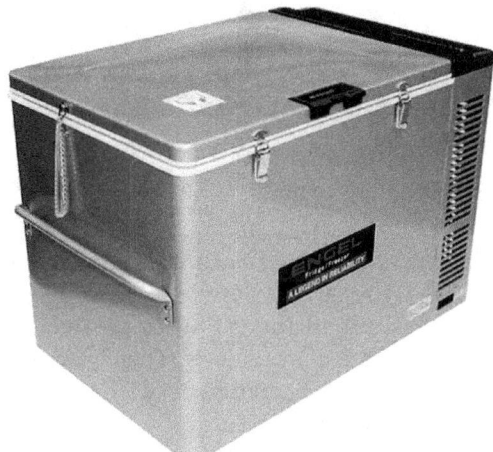

Figure 2.31. Engel 80 litre fridge - efficient and good size for smaller RVs. Pic: Engel.

Refrigerators intended for mobile use (or solar applications) use compressors driven by efficient 12/24 volt motors, enabling them to be run from solar energy. These refrigerators are obtainable from specialised suppliers. The best-known RV brand is Engel (Figure 2.31). It has many excellent rivals.

The Engel's compressor is simple and effective. A permanent magnet at one end of a piston rod interacts with an electro-magnetic coil, resulting in it reciprocating at 50 times a second (Figure 3.31).

The piston assembly's movement (and overall efficiency) is considerably aided by a spring that causes it to resonate at that spring's natural frequency. (Kangaroos' hind leg muscles work in a similar energy-saving bound and rebound action.)

Most other such fridges use one of the wide range of Danfoss compressor motors - Figure 4.31.

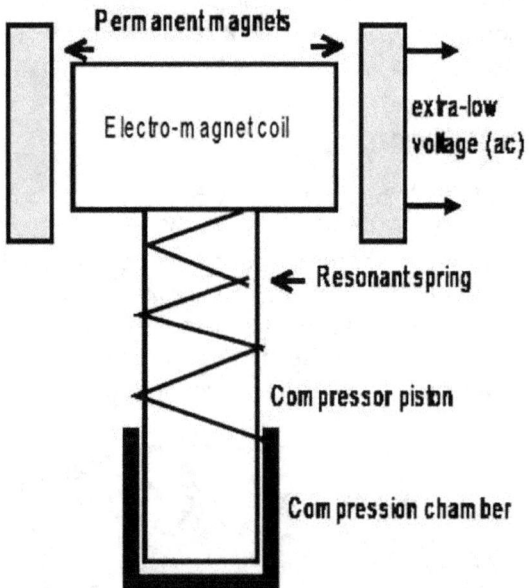

Figure 3.31. Engel swing motor: simple and effective.
Pic: rvbooks.com.au

Until recently the Danfoss types have been used in a cyclic on/off action. They ran until the fridge cooled to a preset temperature. A thermostat switch then cut off the power, restoring it when the internal temperature rose a degree or two above the preset level. They typically cycled in a 40.60 on/off ratio, in ambient temperatures of 25°C or so, increasing to on almost constantly in very hot areas.

There is, however, an increasing shift to employing a modified (and more efficient) form of the motor that runs continuously at whatever speed is required to maintain the required temperature.

Assessing energy usage

Assessing energy consumption is confused by many vendors quoting steady-state draw. This misleads because ongoing draw is also related to the length of time that the fridge is cycled on as opposed to off (and that is related to the thickness and effectiveness of its insulation, etc). A fridge that draws 1.25 amps but cycles on for a total of 16 hours a day uses 20 Ah/day. Another that draws 1.5 amps but is cycled on for 12 hours a day, uses only 18 Ah/day. Consider only the average daily draw.

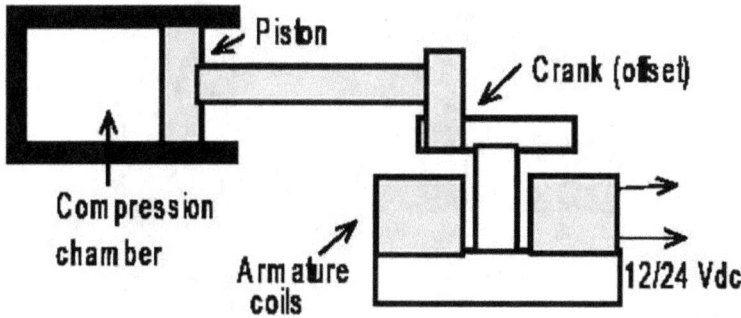

Figure 4.31. Basic Danfoss compressor motor.

At 25°C ambient temperature, and set to 4°C, compressor fridges up to 80 to 100 litre draw about 0.7 Ah/day per litre, dropping to about 0.5 Ah/litre/day as size increases. Fridge freezers, with the freezer set at -14°C -18°C, draw 10 to 15% more, increasing by about 5% per every degree C higher in ambient temperature and by the same amount by which the fridge or freezer is set colder.

Energy draw varies from brand to brand but installation, ambient, set temperature and usage all affect consumption.

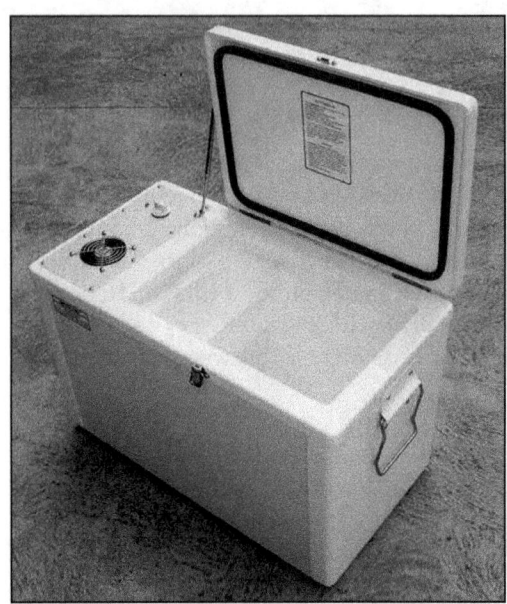

Figure 5.31. Australian designed and built Autofridge. Pic: Autofridge.

The Autofridge (Figure 5.31) has a Danfoss compressor that cools a mixture of salt and water until it becomes solid ice. This results in a so-called 'phase-change', where that ice takes several hours to almost suddenly revert to liquid. These fridges initially need 10 hours or so to drop to about 0°C but thereon need running only an hour or two each night and morning.

At 32°C, both use about 24 Ah/day as a fridge, and 40 Ah/day if used as a freezer (the bigger one has thicker insulation).

Many RV users prefer a 12 or 24 volt compressor fridge if travelling extensively and spending only a day or two on-site. If finance and space for solar modules are not constraints, an energy efficient extra-low voltage fridge driven by the latest Danfoss compressor can be used successfully for extended periods away from mains power but, without that, a fuel cell or generator is likely to be needed for back-up.

Top versus front opening

Top-opening refrigerators are marginally more efficient than door-opening units because cold air is not lost when opened. The amount of cold air lost from door-opening fridges, however, is not great. It can be minimised by using plastic drawers that block the cold air flow. A bigger problem can be that of the magnetic rubber door seals eventually leaking, or the door not being held securely closed. If either happens, energy consumption soars.

One minor drawback of chest fridges is that water vapour condenses in the bottom of the chest and needs removing every week or so. Another is that Murphy's Law of 'Selective Cryogenic Migration' ensures that the most often needed items in a chest fridge invariably shift to the least accessible areas.

Freezers

Freezers use more energy than refrigerators, particularly if freezing content from ambient temperature. They have thicker and/or more effective heat insulation. Once their contents are cold, their energy draw is little higher than a fridge of similar capacity. Because freezers are better insulated, energy can be saved by using a chest freezer as a fridge (set to the +4°C advised).

Fridge standards

Unlike domestic units there are no formal performance standards for RV fridges. European three-way fridges, however, must meet EU Standards that include 'Climate Classes'. This defines the temperature range over which the fridge must deliver its claimed performance. Climate Class SN (sub normal), is from 14°C to 32°C, N (normal) from 18°C to 32°C, ST, (sub tropical) from 18°C to 36°C, and T (tropical) to 43°C. Dometic states some of its range from 120 litres upward meet this standard.

All Dometic RV fridges are 'tropicalised' but that does not imply accordance with Climate Class T. Regardless of brand, Climate Class 'T' fridges can *only* be positively identified by the letter 'T' following the 'Climate Class' box on their compliance/rating plate (normally within the fridge). Figure 6.31 shows an actual example.

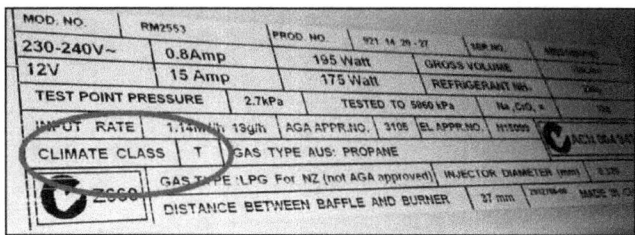

Figure 6.31. A label inside three-way fridges indicates it is Climate Class. This one is 'T'-rated. Pic: Dometic.

The end result

The energy a refrigerator draws depends on several factors. Some are under user control, others not. Avoiding putting in hot items. Do not set internal temperature colder than the recommended 4°C for the fridge and -14°C to -18°C for the freezer.

There is a lot you can do to improve a fridge's installation, and decrease its energy draw. That draw is inversely proportional to a fridge's insulation area and thickness. And, because surface area decreases proportionally with size, always have one large fridge - *never* two small ones - each of half the size. A few quick sums may surprise).

Given adequate insulation, once the contents have cooled, the overall energy used has more to do with the weight and initial temperature of that being cooled, than the fridge's volume. Heat losses through gaps in the insulation and door seals, etc., hugely increases energy drawn. See Chapter 32.

Chapter 32

Installing & optimising fridges

All fridges need to be installed correctly or they will draw excess energy in an attempt to maintain their set temperature. This is essential with three-way units. In practice, very few RV fridges are installed such that they deliver anything like their potential capability. As a direct result, many inherently good fridges may acquire poor reputations through no fault of their makers.

Fridge Capacity (litres)	Vent area (cm^2)
Less than 100	32.0
101-200	45.0
> 200	65.0

Table 1.32. Fridge vent area. (The original is Table 6 of AS/NZS 5601.2010.

The main problems include buying a fridge that was not designed for the conditions/temperature, inadequate wiring and/or ventilation and unrealistic user expectation.

To address these problems, first locate the unit out of direct sunlight. Adequate and effective ventilation is essential, especially for three-way units.

Figure 1.32 (later in this Chapter) shows how a fridge must be installed if it is to perform as its maker intended.

Cool air must enter around floor level. Where there are external cooling fins, baffles should be fabricated and installed to ensure that cool air is directed such that it flows upward through the cooling fins. Many vents are incorrectly located such that the incoming cool air by-passes those fins.

Hot air rises and must be free to exit. The exit vent should ideally be at roof level, or at least 200 mm or so above the uppermost cooling fin. Ducting may be needed to ensure there are no pockets of trapped hot air.

For three-way fridges, AS/NZS 5601.2: 2012 specifies the legally obligatory minimum area of each exit vent. Similar principles (but not legal obligations) apply also to chest-type fridges: cool air must be available at floor level and warmed air must be readily able to escape to atmosphere.

To ensure efficiency limit voltage drop (for 12 volt systems) to 0.2 volt of that across the battery, or better still 0.1-0.15 volt.

Current draw is usually shown in the maker's specifications. Where it is not, assume 40-110 litre chest opening electric units draw about 0.25 amp per 10 litres. Door-opening units draw a little more.

Larger fridges (such as Vitrifrigo and Frostbite) naturally use yet more. Converted domestic fridges (BP, Fisher Paykel, etc.,) may draw 5.0 to 10 amps.

The smaller three-way fridges draw 12 to 15 amps. Those over 170 litres or so draw 15 to 20 amps; 300 litre plus units may draw 25 to 30 amps. This is not a major problem on-road but may become so in the future if the trend to reducing alternator output continues.

For conventional electric fridges, daily draw is a function of their 'on' time versus their 'off' time. A fridge that draws more current, but cycles 'on' less frequently or for a shorter time may thus draw less energy per day than a unit with lower energy draw that is cycled 'on' more often, or for longer. The energy draw of the more recent variable speed motor units is shown in their makers' data.

Take fridge wiring voltage drop very seriously. Because copper is expensive, few RV installed fridges are likely to have adequately rated cable. Almost all can have their energy usage reduced, and cooling improved by increasing cable size. Often dramatically so.

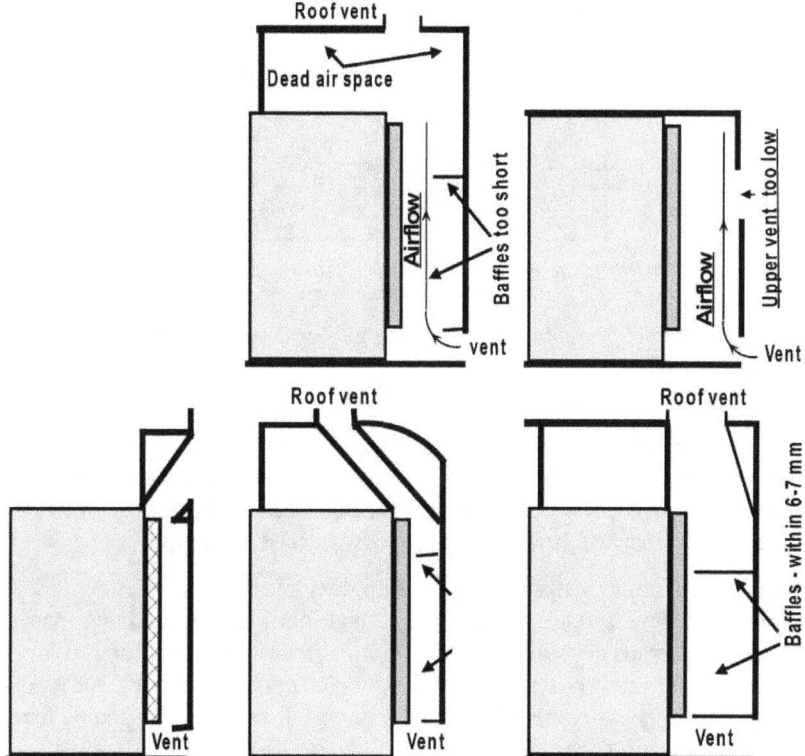

Figure 1.32. How to install a fridge correctly. Cold air needs to enter from below the cooling fins and be directed by baffles to flow only through those fins. Baffles must extend to within a few millimetres of the fins. Hot air must be channelled to the exit vent.
Top left: *too short baffles cause incoming cool air to bypass the cooling fins; rising hot air is trapped, hindering air movement. Top right: the upper vent is too low, trapping hot air above.*
Bottom: *here's how it should be done. An extractor fan helps shift the hot air but is not a substitute for a proper baffle. Drawing: 2018rvbooks.com.au*

Never set the thermostat to a lower setting than necessary. Energy consumption increases by 5% for every 1°C below the recommended 4°C for the fridge, and -14°C to -18°C for the freezer.

Electric-only units up to about 220 litres can practically be run from solar. Older larger units draw too much for (RV) solar to make economic sense, but many top quality post-2014 are extremely efficient if competently installed. Some of about 300 litre capacity, draw under 1000 Wh/day, and are fine for motorhomes and fifth wheelers that have sufficient roof space and battery carrying capacity.

Reducing heat loss

Litres	50 mm	75 mm	150 mm
30	210	140	70
60	320	213	107
120	420	280	140
240	637	425	213
480	840	560	280

Table 2.32. Typical heat loss (in Wh/day) for varying insulation thicknesses.

As a refrigerator's interior air and contents cool, energy is expended in pumping out heat that has found its way in through inadequate or ineffective heat insulation (Table 2.32).

More energy is expended while opening and closing the door as cold air falls out, and warm air gets in. This can be reduced by adding light plastic drawers. Further energy is lost through poor door seals. Energy draw depends substantially on how well these losses are stemmed.

By far the best heat insulator is a vacuum. Dry motionless air is a good second best. The latter is exploited by a high-density polyurethane foam material that traps air within its honeycomb structure - ideal for refrigerator construction. It costs more than light-density equivalents - but its extra cost is negligible compared with the energy it saves. It is available from Clark Rubber and other suppliers.

Heat loss is inversely proportional to insulation thickness, and directly proportional to its area, so a refrigerator's shape is important. The further away from a globe (realistically a cube), the greater the surface area and the greater the heat energy lost.

Domestic fridges in RVs

There is a growing trend to use top quality domestic 230 volt fridges (driven from 12/24 volts via an inverter). There are no problem with ruggedness - not least as many sold as RV fridges are modified domestic units anyway. This is worth investigating as the top quality post-2014 domestic fridges are extremely efficient and a few made subsequently are even more so.

Be aware that, for reasons that defy sanity, a typical caravan door opening is 20 mm less wide than a fridge's typical 600 mm so the fridges are usually installed before adding the door frame. Removing and replacing the frame however is usually feasible (Mielle make a good fridge that is 580 mm wide).

Some of these fridges have the condenser fins inside the fridge's metal walls - these fins are used to radiate the heat. This precludes adding insulation.

An air space of 50 mm must be left on either side and to the rear of such fridges - and ample space above to allow rising heat to escape. A small electric extractor fan usefully assists.

Building your own fridge

Building your own fridge is relatively easy. Doing so can make a great deal of sense as it can often be designed to the size and shape that you wish, yet commensurate with efficient operation.

This particularly makes sense for those who fish seriously. Cooling and freezing fish needs a lot of energy but there need be only minor energy losses thereafter, even with large capacity storage.

As the table (2.32) shows, tripling insulation thickness (from 50 mm to 150 mm) results in the heat loss of a 480 litre fridge being less than that of a 60 litre fridge.

Building your own fridge is eased by DIY kits supplied by companies including Engel, SEA FROST (USA) and Waeco. All usefully enable you to locate the compressor remotely from the actual fridge compartment. Some commercial fridges too (e.g. Vitrifrigo) have a remote compressor.

Such fridges have the advantage that extra insulation can be added all round. Another benefit, particularly for off-road use, is that no external vents are needed within the RV's living area.

The SEA FROST product (Figure 2.32) is used extensively by sailors (for whom building a fridge is almost routine). That, and similar systems, consist of a compressor and a stainless steel cold plate.

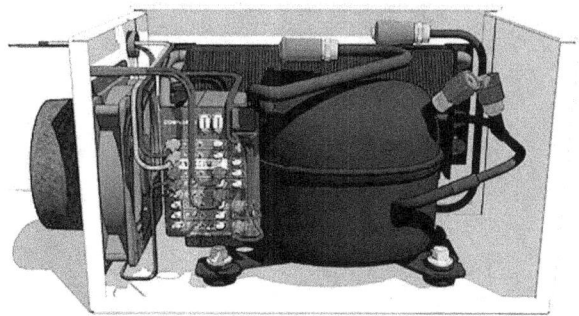

Figure 2.32. Typical SEA FROST compressor section is roughly 200 by 200 by 400 mm, and weighs 9 kg.

The working bits must be located such that cold air can be drawn in without restriction, and rising hot air vented to the atmosphere, perhaps assisted by a small extractor fan.

Those sold by Jaycar Electronics and others for computer cooling are ideal. Dometic supplies an excellent DIY roof vent kit.

Installation of a SEA FROST system is eased by it taking in fan-drawn outside cool air and exhausting the warm air to the atmosphere, both via 100 mm flexible tubing.

For fridges, the cooling plates typically take up about 20% of the inside surface area. For fridge-freezers, that area may be about 60% of the sides. Fridges and fridge-freezers need different thermostats.

As noted previously, cooling efficiency depends on the quality and thickness of the insulation - and is best achieved if the fridge is made as a top opening chest unit.

For the SEA Frost unit described above, the manufacturer states that given 77 to 100 mm thick top-quality insulation, the following can be achieved in tropical climates: 'fridge capacity 65 litre: current consumption will be approximately 25 Ah/day. If capacity is 225 litre: current consumption will be approximately 57 Ah/day. If capacity is 440 litre: the current consumption will be approximately 70 Ah/day.' If that insulation were to be 150 mm, energy usage would be even less.

The ideal might well be one small fridge/freezer for rapid freezing, and a larger one for storage once the fish is frozen.

Chapter 33

Television

Whilst less so with digital television (than the previous analogue), TV reception in an RV may still involve compromise. Whereas a fixed location enables an optimum antenna for your specific location, an RV-based antenna must be able to receive signals, depending on where it is, from any of many transmitters. These may be VHF (Very High Frequency) and/or UHF (Ultra High Frequency).

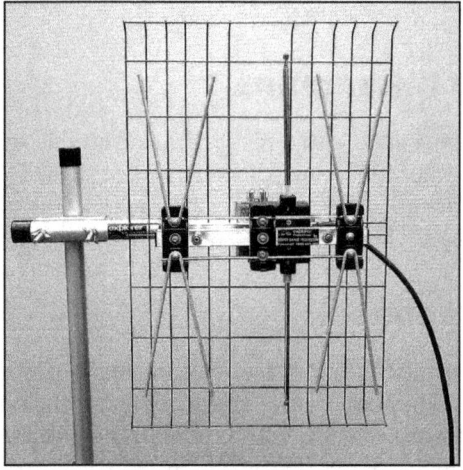

Figure 1.33. The Explorer C2sg UHV/VHF antenna. Pic: Explorer.

This choice of vertical or horizontal polarisation reduces interference between TV stations with overlapping geographic coverage. City-based transmitters variously need the receiving antenna to be horizontal or vertical. An antenna that can do this is inevitably a compromise between size, signal gaining ability, directionality and ability to receive all channels.

A typical antenna picks up transmissions in a flattened balloon pattern. The non-mast end usually points directly toward the transmitter.

The signal is however compromised by distance, obstructions and reflections from buildings, hills, etc. In some locations, the strongest signal can be that reflected from a nearby building or a hill.

The lower the channel number, the larger the antenna elements required: from 2.6 m wide for channels 1 to 2, to 350 mm for channels above 40.

VHF/UHF antennas

Most regional areas use smaller transmitters that use UHF frequencies. These also fill 'holes' in city areas. Given a good antenna, signals may be receivable from 70 to 100 km away. Small country town stations are UHF with fringe range of 40-60 km.

For VHF/UHF a good choice is a so-called phased array antenna. That shown in Figure 1.33 is the Explorer C2sg. It has excellent fringe area reception. The smaller C3sg is equally effective with UHF but the makers say it is less effective with VHF.

Dome type omni-directional antennas are compact and convenient. They are fine close to towns but many users claim that, generally, they do not provide as reliable reception as those noted above.

Connecting the antenna

Connect the antenna via the shortest possible route using low-loss coaxial cable. People are reluctant to perforate roofs to take a TV cable but waterproof glands (that keep water out of ocean-racing yachts) are available from boat chandlers. Through-wall connection kits are also available.

Antenna amplifiers

Antenna amplifiers (also called 'mast head amplifiers) assist in electrically-noisy fringe areas. They reduce the effect of interference picked up by the cable from the antenna to the TV, and may also be effective in fringe areas by enabling a 'clean' but weak signal to be processed correctly. This is a common need in an RV. Have the antenna amplifier as close as feasible to the antenna.

Whilst rarely used nowadays, the set top boxes used to enable analogue TVs to display digital programs need clean 230 volt power. They may not work satisfactorily, or may be damaged, if run from anything other than mains power or a pure sine-wave inverter. As the total energy used far exceeds that of the latest digital TVs it makes every sense to scrap the existing set up and buy anew.

Satellite television

Many RV owners use satellite reception. Its major benefit is that programmes can be received from any number of sources worldwide, including both paid TV and free-to-air.

As with digital transmission of any nature, there is either a virtually perfect signal (e.g. picture) or none at all.

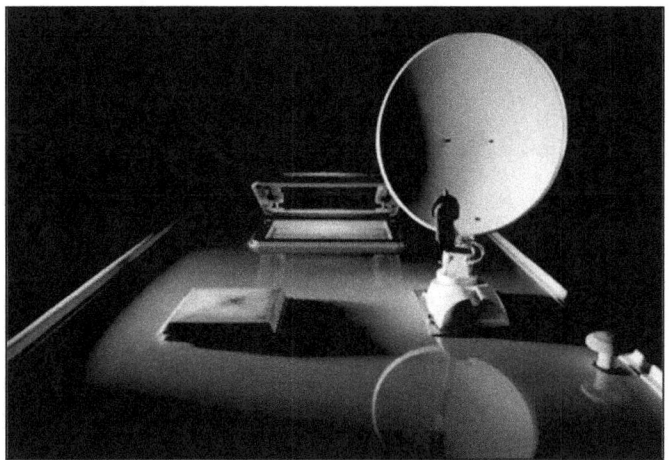

*Figure 2.32. Oyster Vision automatic tracking satellite TV antenna.
Pic: southdowns motorcaravans.*

A 650 to 800 mm diameter dish antenna will provide good signals in clear sky conditions but signals may drop out in heavy rain. Bigger is better.

There must be a clear 'line of sight' to the satellite, and the antenna must point directly toward the satellite. This is reasonably easy to do manually given a $20 to $100 satellite tracker. The more upmarket systems have automatic tracking (Figure 2.32). Installation must allow for the satellite dish to tilt and swivel. If used in varying parts of Australia, this requires a range of 30°C to 70°C from horizontal.

Power consumption (of the antenna system) is typically 50 watts while searching for the satellite, and 10 to 15 watts thereafter. Weight is typically 10 to 15 kg.

All but the most basic of RV antennas really need professional installation. If you do it yourself, follow the maker's instructions to the letter. Do not skimp on the coaxial connecting cable. High performance cable is costly but well worth the price.

People travelling in caravans or motorhomes in remote Central and Eastern Australia are eligible to apply for temporary access to Commercial TV on VAST (Viewer Access Satellite TV).

Satellite pay-TV reception is provided by the Optus C1/D3 satellites and receivable Australia-wide. These satellites also carry the VAST free digital TV service that provides a full quota of standard definition and high definition free to air channels plus a wide range of local news channels for many areas of Australia. Travellers are required to register before the system (remotely) switches on the decoder.

Apart from a suitable dish you need a 'low noise block converter' (about $30 and usually supplied with the dish), a VAST digital satellite receiver with included Smartcard. There is no licence fee.

For further information on VAST commercial services see mysattv.com.au.

Television receivers

Well over half a TV set's energy is used by the screen and its drivers, and by an amount more or less proportional to the square of the screen width. LED/LCD 26 to 35 cm (10 to 14 inch) TVs use 20 to 40 watts. Plasma screen TVs are best replaced by LED/LCD units as they draw far less energy.

Any suitable fringe area LCD or LED TV is suitable for RV use. Most owners use a domestic 230 volt unit run via an inverter. These are generally cheaper and the best ones use far less energy. If the RV already has an inverter, there is no point in buying a 12 volt unit as they are often not that efficient 230 volt or 110 volt units with a small inverter inbuilt anyway. This handy if you have no inverter, but if you have, settle for a 230 volt TV.

For RVs that have no inverter, 12 volt TVs are available with screen sizes from 18 cm (7 inches) to 80 cm (32 inches).

A lowish energy solution is to use a recently made laptop computer (Figure 3.32). Some have a DVD player inbuilt. Some also have an inbuilt TV tuner and have excellent picture quality.

Figure 3.32. Some laptop computers double as a TV - and as does this one, have an inbuilt TV tuner. Pic: TVPCSAT

Standby consumption

TVs and many domestic appliances need to be turned off at the wall switch, not just by the remote control. Most post-2014 high quality units draw only one watt when switched off at the remote, but older appliances may draw up to 15 watts (and thus consume most of their energy when not in use!).

Computers and TVs withstand normal road shocks, but not unrestrained jolting. They should either be located securely or packed so that they cannot move.

Rough roads and corrugation, however, may well damage rigidly located TVs and computers. Sprung mounting assists but must be very securely fastened, as ongoing corrugation may cause the bracket to vibrate and fracture, or the screws to be pulled out of the wall. Then down comes bracket, TV and all. Unless truly secure mounting can be assured, it is better to pack them bubble wrapped in a secure dust-proof box while travelling.

Chapter 34

Communications

Telstra's Next G, 4G (and soon) 5G services increasingly provide coverage across most of the more populated areas of Australia, and along many of its major highways. As of mid 2019 most of the more populated areas of the Northern Territory and inland Western Australia are covered, but much of the land mass is not, nor is likely to be excepting for areas around Aboriginal communities, major cattle stations and a few of the larger mines, etc.

Figure 1.34. 5G USB modem. Pic: Telstra.

The Next G and 5G mobile phone bands can also carry email and internet traffic on mobiles, smartphones and iPads/tablets. Laptop computers and iPads/tablets connect to Next G and 5G using either a USB modem (Figure 1.34), a Wi-Fi modem or a built-in SIM card. The new smartphones can be used as 'Wi-Fi hotspots' to connect other devices, such as laptops and iPads, to the internet. This enables using the included mobile phone data allowance or additional browsing-pack data.

Some older modems require a 230 volt supply. They are readily run from solar via a pure sine-wave inverter of 50 watts but it is worth buying one that is larger as it will prove useful for powering other mains electrical equipment. Newer wi-fi modems have inbuilt batteries that need to be recharged via a 230 or 12 volt powered charger.

Mains-voltage phone chargers draw only a few watts but will keep an inverter drawing continuous power once the phone is charged. Older mobiles recharge in an hour or two. Smartphones take a bit longer but the inverter and power supply continue to use energy for the rest of the night. USB adaptors that plug into 12 volt cigarette sockets are also readily available. Some RVs have USB sockets.

Figure 2.34. Smartphone installed in a motorhome. Pic: Laurie Hoffman.

In-vehicle mobile phone kits have efficient inbuilt 12 volt chargers. Some connect the phone to an external antenna, considerably enhancing signal strength, and extending the phone's usable range.

Satellite communications

These telephones are ideal for outback travelling in Australia. Across much of the interior they are the *only* possible method of such communication. Their major limitation is that their (still 2019) slow speed of 2400-9600 bits/second effectively limits their use to voice and plain text emails.

The handsets resemble mobile phones (but are thicker and have a short antenna). They work in a similar way, excepting that some services require users to include full country and area codes, even for local calls.

All require a more or less unobstructed line of sight to the satellite, so these phones rarely work inside homes or vehicles: they need an external antenna.

There are several and different satellite services, of which some have only limited coverage. For use in Australia and much of Asia, the Thuyara service provides adequate coverage, as does Globalstar.

The Inmarsat service, however, is truly global, excepting, (still in June 2018 at least), to North Korea - for current 'political reasons'.

Most services have a monthly fee comparable to mobile phones but call charges are substantially higher. The Vodaphone system first checks for a

Next G signal and if that is not obtainable, only then switches to the far more costly satellite operation.

Satellite handsets

Most hand-held mobile satellite phones (Figure 3.34) have a short extending antenna - but some have provision for plugging in a high gain antenna for use in areas of marginal reception.

Figure 3.34. The popular Iridium 9555 satellite phone. Pic: Iridium.

Satellite phones use more energy than do normal mobiles but can be powered from 230 volt inverter power, or alternatively via 12 volts dc or a USB port. They draw only minor power and can readily be run from solar.

High frequency (HF) radio

The HF radio service provides communications coverage only between HF transceivers across limited and far from predictable areas. Also, there is a 'skip distance' of typically 70 to 100 km from the transceiver across which communication is only rarely possible.

Apart from specialised use, HF radio is best left to its band of enthusiasts, for whom it is part communications but mostly hobby. It is also a heavy power user - although solar can cope.

UHF CB radio

CB radio is now extended from the previous 40 channels to 80 channels. A high gain antenna extends its range but some users prefer a short half-wave one that provides better clarity over 5-10 km. Range depends also on the nature of the terrain. It may vary from line of sight to 50 km in rare conditions.

Short range handheld transceivers are also available. Caravanners in particular use them to advise their partners when reversing into tight places, etc. This can be done by having a pair of transceivers, or having just one used in conjunction with the vehicle's own CB - (and is cheaper than a divorce).

In-car CB radios are normally muted to remain silent between signals. It is easy to forget they are still on when you leave the vehicle, resulting in a flat starter battery (but readily recharged if you have solar).

Useful links

telstra.com.au/mobile-phones/coverage-networks/our-coverage/coverage-search/

telstra.com.au/mobile/networks/coverage/broadband.html

telstra.com.au/bigpond-internet/mobile-broadband/

https://telstra.com.au/mobile-phones/nextg-network/

Chapter 35

Electrical & radio interference

Hums, whistles and crackles on radio, HF or CB, are mostly caused by radio frequency interference (RFI), and via wiring from the alternator and ignition system. Clicks are caused by turn indicators, crackles, and poor connections and whines from electric motors in windscreen wipers, heaters and air conditioners.

If the source is hard to locate, you can readily find it by using a portable analogue radio (they have very directional inbuilt antennas). Turn it around from two places about 45 degrees apart such that the RV noise is maximised and at each, draw a line through the centre of the radio toward the source of noise. The source is likely to be where the two lines intersect.

A healthy ignition system may cause slight interference in fringe reception areas but, if faulty, will generate radio frequency noise in most areas. It often originates from excess plug sparking voltage caused by worn or incorrectly gapped plugs, slight breaks in plug or coil leads, excess gap between rotor and distributor segments, or wrong ignition coil polarity.

Because electrons need less voltage to jump from hot to cold surfaces than vice versa, ignition coils are wired so that the spark jumps from the centre (hot) electrode to the colder plug body. If that polarity is correct, 15,000 to 20,000 volts suffices but 30,000 volts may be needed if not.

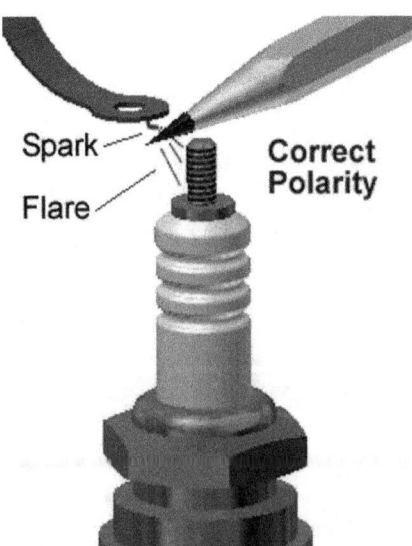

Figure 1.35. How to check coil polarity.

To check coil polarity, disconnect a plug lead, hold a pencil tip slightly away from the plug terminal, and the plug lead an equal distance on the other side of the pencil - Figure 1.35. Correct polarity causes an orange flare on the plug side of the pencil.

If the polarity is incorrect, interchange ignition and distributor leads on the coil regardless of their marking. (Coils intended for positive earth ignition are identical except for that marking).

To further quieten, connect a 1.0 uf (micro-farad) capacitor between the ignition switch positive and earth. Radio interference may also be radiated from the cable between tachometer and alternator. Its metallic braid must be well earthed at both ends.

Plug and distributor/coil leads must be sound and kept away from other leads. Plugs need to be correctly gapped, and with electrode edges sharp.

Apart from RF noise, incorrect polarity also affects starting, and may cause erratic running at light throttle because lean mixtures require a higher voltage to ignite.

If a whine varying in pitch with engine speed is generated by the alternator, try a paralleled alternator-to-battery lead. If that helps, replace the original lead by light starter motor cable. It that too does not work, connect a 1.0 uF 230 volt capacitor between alternator output and earth. (Jaycar catalogue number AA 3060). Do not connect a capacitor between the alternator field and earth as it may affect the voltage regulator.

Windscreen wiper motors are quietened by connecting a 1.0 uF capacitor between each brush and earth, and maybe an additional 0.001 uF capacitor across the input, close to the wiper motor. Clicking switches are fixed the same way.

Chapter 36

Lightning protection

The earth's surface has a negative electrical charge, the upper atmosphere a positive electrical charge. The resultant difference, of 10-100 million volts, can result in an initial ionizing arc of about 1000 amps, then a major discharge of up 100,000 amps for a millisecond or so.

In most of Australia the probability of an RV being struck is slight but it is higher in the Blue Mountains, the Dandenong Ranges, the northern parts of North Queensland, the Northern Territory, and parts of Western Australia (where a full-on Kimberley storm is quite an experience). The statistical risk is still slight but as storms may be intense, some people tend to feel safer if their RV is protected.

Aluminium or steel panelling provides excellent protection: it acts as a so-called 'Faraday cage'. Electrical discharges are conducted to earth via the metal exterior of such cages. Major metal structures (e.g. air conditioners and metal frames of solar modules) both inside and outside the vehicle, should be earthed to the vehicle's chassis by 6 to 8 mm² cable. Tyres, now partially conductive to reduce static build-up, route lightning current to earth.

Within such vehicles, occupants are not likely to be harmed or even aware of a strike providing they stay away from windows, walls and any major interior metal such as a fridge. Any exterior power cord to a generator or caravan park supply is best kept disconnected and kept well away from the RV.

Fibreglass bodied RVs can be partially protected by a lightning conductor - an earthed elevated rod that theoretically provides a 'cone of protection', of which the diameter at ground level is about the height of the spike. While once believed to assist to dissipate the electric charge, it is now generally believed such 'protection' marginally increases the possibility of a strike but minimises the effects.

Campfire and forum mythology often suggest that corner jacks be insulated via thick timber blocks but overlooks that a strike that extends a kilometre or more is hardly going to have a problem bridging a further few centimetres of timber block to the ground. Doing so also introduces the risk that people stepping out of the RV, and thereby momentarily forming a conductive path to earth through themselves (in parallel with the tyres), might initiate a strike. In practice, if you can hear thunder or see lightning, stay inside. Your distance from a witnessed strike is 1 km for every three seconds delay in then hearing it.

Figure 1.36. Lightning strike over Geelong (Vic).
Pic: (by Rod Howard) courtesy Geelong Advertiser.

Chapter 37

Caravan specific issues

A conventional caravan's electrics are substantially similar to those in camper vans and motorhomes except for one vital and major difference: their batteries are a lot further from the vehicle's charging source. A fifth wheel caravan's electrics are likewise.

Delusions of adequacy

There are no Australian standards for voltage drop in RV wiring, nor for RV wiring generally (excepting for front, rear and stop lights and braking). One cannot therefore assume that all is fine because it is a new rig. As auto electricians confirm, some are not.

Many caravans systems lose a volt or more between the alternator and the remote located battery while attempting to charge it. As a result, unless a dc-dc alternator charger (Chapter 13) is installed, few caravan-located batteries reach 70% of full-charge, and some less.

Voltage drop results in caravan fridges failing to cool adequately and/or drawing excess energy. This is particularly so with three-way fridges. These draw 15 to 25 amps from an intended 12.6 volts while driving but are often so hampered by voltage drop that they barely cool at all. The major cause is that the desired cable is a lot thicker than most installers use. That is also costly and too thick to fit into standard 7 or 12 pin connectors. Most vans are thus wired with cable rated at a fraction of that required for minimal voltage drop.

For a 12 to 14 metre conductor run in any charging circuit, the minimum acceptable size for direct alternator charging is 10 mm^2, but heavier is preferable. This enables reasonably effective charging for close to flat 100 A/h batteries.

The above will work even better with AGM and lithium batteries as these charge faster and more deeply at a lower voltage. The only *truly* effective solution is to install an adequately rated dc-dc alternator charger or battery management system in the caravan and close to the battery bank.

You still need thick cable (10 mm^2 is a realistic compromise but 13.5 mm^2 is better) and a genuine Anderson plug and socket between the vehicle and trailer. This is costly but it works.

Trailer connectors - Australia

Vehicles and trailers fitted with electrical connectors between 1982 to 1985 are likely to have a 7 pin device manufactured to Australian Standard AS 2513.

Connectors that met this standard were made by several local companies. Identical-looking connectors (that did not comply with AS 2513) were also imported from Asia.

The AS 2513 standard covered useful things like current-carrying capacity, corrosion resistance, etc., but neither size nor shape. Some AS 2513 plugs and sockets were flat, others were round. To assist compatibility, flat/round adaptors were made by Utilux, Brylite and others. They were/are handy for temporary situations but it is never a good idea to have multiple connectors.

Semi-sanity set in around 1995 with the introduction of AS 4177-5-1995. This specifies flat 12 pin arrangements (only) but the AS 4177-5 socket accepts flat 7 pin and 12 pin AS 2153 plugs. There is now also a 13 pin plug and socket.

Brylite produces separate four-pin, 35 amp per pin plugs and sockets specifically intended for loads such as a three-way refrigerator, and for battery charging but the Anderson plug/socket is more generally used in Australia and New Zealand.

Pin numbering

Figures 1.37 and 2.37 below show recommended connections for round 7-pin sockets, plus 7-pin and 12-pin flat Brylite sockets - both as seen from the rear (wiring) side. With all of the above sockets, pins 1 to 7 are rated for 15 amps for one hour. Pins 8 to 12 of the 12 way socket are rated for 35 amps.

Connector sockets have a cover for RV non-use, but not for plugs. As a result, the plug tarnishes, causing voltage drop. Clean the plug or fit a new one, and make up a waterproof protector.

Table 11: 7/12 pin connections

1. Left indicator	Yellow	7. Rear, clear, side markers	Brown
2. Reversing signal	Black	8. Charging/winch	Orange
3. Earth	White	9. Auxiliary circuits	Pink
4. Right indicator	Green	10. Earth return	White
5. Electric brakes	Blue	11. Rear fog light	Grey
6. Stop lights	Red	12. Spare	Violet

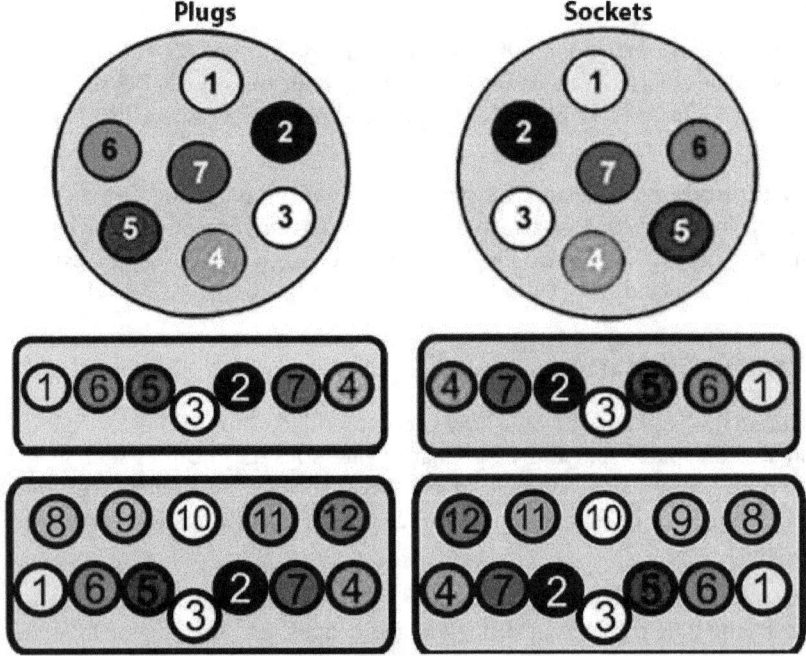

Figure 1.37. This is the standard pin numbering and cable colour - seen from the wiring side - as set out in VSB-1. Seven-pin plugs (only) confirm to Australian Standard AS 2513. The current AS 4177-5-1995 standard relates to the 12-pin arrangement (only). See main text re industry use of pin 2 of 7-pin connectors. Contrary to general wiring convention, white is used for chassis and earth. Some Australian states used varying connections for early 7-pin connectors. For a colour version of this image see the back cover of this book.

Trailer connectors (13-pin) - Europe

Since 2008, European caravans have used a 13-pin plug and socket. The connections are shown in Figure 2.37. As can be seen it has no specific provision for a winch.

Table 12: 13 pin connections

1. Left indicator	Yellow	7. Left tail/side ights	Black
2. Fog light	Blue	8. Reversing light	Pink
3. Earth	White	9. 12 volts battery power	Orange
4. Right indicator	Green	10. 12 volt (ignition on)	Grey
5. Right tail/side light	Brown	11. Earth for 10 (above)	White/Black
6. Brake/stop lights	Red	12. Not normally used	White/Blue
13. Earth for terminal 9	White/Red		

Figure 2.37. European 13-pin plug - as viewed looking into the plug. For a colour version of this image see the back cover of this book.

Electric brakes

An electric drum brake (Figure 3.37) has an electromagnet that, when braking, presses against the revolving drum, causing brake shoes to be thrust against that drum. Only 10-50 watts (1.0 to 6.0 amps) is required for full actuation. Power is normally applied automatically via a brake controller in the tow vehicle.

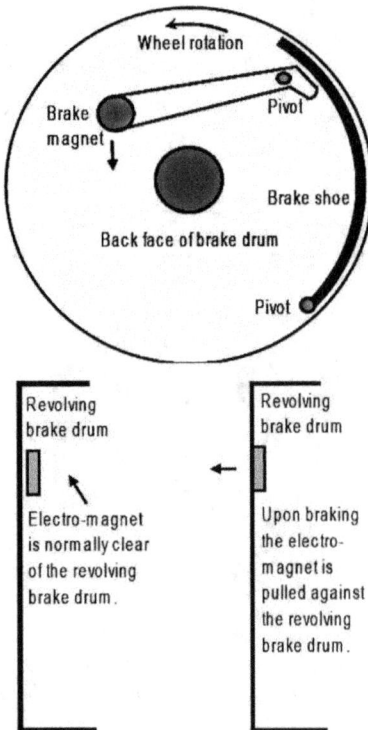

Figure 3.37. When braking, an electromagnet causes a pivoted lever to force the brake lining against the rotating drum's back face. A second brake shoe, on the 'left-hand side' of the drum, is not shown. Pic: rvbooks.com.au

A basic controller is a pendulum free to swing forward. When the vehicle's brakes are applied, its momentum causes the pendulum to swing proportionally to braking force, varying the actuator's electrical resistance, enabling it to actuate the electromagnets in the brake assemblies and apply the brakes.

A manually operated override (that provides back-up) may also be used to operate the trailer brake alone, for example to stabilise the outfit on slippery hills. An adjuster enables the overall system to be fine-tuned to allow for changes in trailer weight, and for the degree of trailer braking required. The trailer's brakes must *never* be set so that the caravan brakes lock up - Chapter 37.

Most brake controllers are more complex than this but the basic principle still applies: they actuate the trailer brakes in proportion to the deceleration of the towing vehicle. The new solid-state units provide more progressive braking.

Electric brake connection

Most brake controllers have four coloured leads. White goes to earth, red to the stop-light switch, black to the positive of the 12 volt supply, and blue to the trailer's brakes. This is usually via pin 5 of a 7 or 12-pin connector.

Australian Standards 2513 and AS 4177-5 recommend that pin 5 be reserved for electric brake connection - but that cannot be relied upon to have been done. One owner cooked the electric brakes within 30 minutes on a new camping trailer when he connected it up to a new 4WD. The maker had used pin 5 to supply 12 volts to the trailer's electrics, and had connected the brakes through pin 9 - that was wired conventionally and correctly to carry a permanently 'on' 12 volts.

In practice, some Australian trailer and caravan manufacturers use pin 2 of 7-pin connectors for auxiliary lighting. This precludes the use of reversing lights. It also appears to contravene ADRs (Australian Design Rules relating to vehicle on-road safety standards).

Braking problems

First, check that the system is adjusted correctly. Procedures vary but you need to follow the manufacturer's instructions to the letter. If this is not successful, and assuming the brakes are mechanically sound, problems are usually traceable to intermittent current flow, or voltage drop introduced by a dirty or corroded trailer connector.

Measure the voltage across the electromagnets, or as close to them as possible, while operating the manual override throughout its full extent. There is often an indicator light that varies in brilliance as you do so. This should result in a voltage that increases smoothly, from zero to a maximum that depends on the controller's setting. Wiggle the connecting plug to see if there's any variation. Do not do this testing for more than a minute or two or the brakes' electromagnets will overheat.

It is possible that the cabling is inadequate but improbable if the brakes were previously working satisfactorily. A 3% (0.36 volt) drop is acceptable as braking is at its maximum well below full voltage. Check also for a loose fuse (although brake circuits are better protected by a circuit breaker). If all is well, the problem lies in the automatic control mechanism. Have a quick check for loose connections, etc., but it is otherwise best to leave this to a specialist in the field.

Random brake operation

A few owners have reported that their trailer brakes sometimes actuate 'of their own accord'. This can often be caused by radio frequency interference from within the RV, or from CB radios or HF transmitters nearby. Wiring a 1.0 uF capacitor (from Jaycar or Altronics) across the incoming power cable close to the brakes usually solves this problem.

Further, while it might seem blindingly obvious, effective braking requires braked wheels to be firmly on the ground at the time. This will not be so on rough surfaces or corrugations unless the suspension and shock absorbers ensure they are.

AL-KO ESC

From 2012 onward, some caravans with AL-KO suspension (only) have AL-KO's Electronic Stability Control. This is an automatic system that senses swaying of such magnitude (about 0.4 g) that it would lead otherwise to jack-knifing. When that level is sensed the electro-mechanical mechanism automatically applies both side's trailer brakes at about 75% of full braking for one to three seconds - repeating if necessary. This assists to straighten the rig and, by reducing speed, likewise reduces the effect of the forces responsible for the unstable situation.

Tuson (Dexter) ESC

This is a US electronic stability control system that works much as above but applies differential trailer wheel braking. It operates at lower sway levels (about 0.2 g). Whilst effectively correcting sway, it is RV Book's opinion that by masking swaying at low levels this masks underlying instability issues in caravan and tow vehicle combinations that need addressing at source.

* For a technical explanation of the above, and why locking trailer wheels whilst braking must be avoided, please see the author's paper:rvbooks.-com.au/caravan-and-tow-vehicle-dynamics/

Do see also our specialised book '*Why Caravans Rollover - and how to avoid it.*'

Chapter 38

Example systems

Typical basic usage

An RV's most common use is weekends, school holidays and occasional longer trips. Owners increasingly free-camp, with stays every few days in caravan parks.

Typical equipment includes a (now) 60 to 80 litre 12 volt fridge, an electric water pump, inside and outside lighting, a portable radio and a smallish TV. In most, the alternator charges a conventional deep-cycle battery via a manually operated isolating switch, or high current relay switched via the ignition, or post-2000 or so, via a voltage-sensing variant of that relay.

The above works fine for a typical overnight stay, and a second night at a pinch, as long as the vehicle is driven for a few hours in between.

By 1990 or so, a few owners began to add solar capacity to assist charging, and help power the fridge. A surprisingly large number of older such vehicles are still in use - particularly Toyota Coaster conversions.

In the early 1990s we rebuilt our 1974 Kombi for extended dirt road use. The vehicle readily coped with well over 50,000 km of major outback tracks including the Birdsville Track. Its later sale has been regretted ever since!

Figure 1.38. Our 1974 Kombi with tilting solar module. Pic: near Birdsville 1995.

Fully off-road (OKA)

Our 1994 fully off-road OKA had an upgraded (140 amp) Bosch alternator with a 1995 TRW multi-stage external regulator. It initially charged three 100 Ah 12 volt paralleled batteries.

We originally fitted a suitcase-sized (1995) Westinghouse satellite phone system that weighed about 15 kg, drew about 200 watts - and proved 100% reliable. Replacing the Westinghouse satellite phone (by 2002 Iridium hand-held unit) enabled battery storage to be reduced to a 135 Ah AGM.

Figure 2.38. The OKA's interior was designed by Maarit Rivers - and reflects her Scandinavian background. Pic: rvbooks.com

The OKA had eight individually switched 10 watt internal halogen lights plus insect-repelling exterior lights and a 71 litre Autofridge. A 12 volt pump provided filtered water from twin 120 litre tanks. It also had a diesel interior heater.

It is are amongst OKA conversions in being only 5.2 tonne with full (400 litre fuel tanks and one 200 litre water tank). This was achieved by building the interior using mainly white powder-coated aluminium on an ultra-light spruce frame. Most such weigh 6-7 tonne.

We used the OKA for ten years across major and minor tracks in Australia, from our then home outside Broome - and crossed Australia via Alice Springs and back over twelve times.

Figure 3.38. The OKA camping in the Kakadu National Park. Pic: rvbooks.com.au

Nissan Patrol/TVan

In 2005 we decided to trial a fully off-road tow vehicle and camper trailer. We bought one of the last made 4.2 litre TD Nissan Patrols, and a then recently-introduced 750 kg TVan.

Figure 4.38. Not bogged! - dropping tyre pressures to ease going over 40 km of soft sand on part of the Canning Stock Route (northern WA) en-route on one of our routine trips from Broome to Sydney and back. Pic: rvbooks.com.au

Knowing it was feasible to run the electrics from solar alone, we installed two independent systems. A 60 litre Engel fridge in the Patrol was powered with ease by a 120 watt Kyocera solar module mounted on roof bars, and a 110 Ah Ritar AGM located directly behind the steel cargo barrier. The Iridium satphone was charged by the Nissan's 12 volt system. A 25 mm² cable plus connectors could parallel the auxiliary and starter batteries if needed but was used only occasionally - to jump start our big 4WD Kubota tractor on our then Broome property.

The Tvan's original halogen lights were replaced by LEDs and its original twin water pumps by one quiet Shurflo Whisper King unit. We added a Webasto diesel-powered hot water and space heater, 12 volt electric blankets, and a 250 watt inverter for our laptop computer and Next G modem. All were handled with ease by a single 60 watt Kyocera solar module, a Plasmatronic PL 20 regulator and a 12 volt 110 Ah gel-cell battery.

This rig was used for three further 'across Oz and backs' - including via the remote Talawana track (that includes part of the Canning Stock Route). It was 100% reliable.

A 24 volt system

A Nissan Civilian coach-based conversion owned by a friend has a 24 volt alternator and starter batteries. It originally had an 80 litre fridge/freezer, 12 volt TV/DVD player, three 35 watt halogen lights plus two similar reading lights and a water pump. All now run from two 100 Ah, 12 volt parallel-connected gel-cell batteries charged by the Nissan's alternator via a Redarc 24-12 volt dc-dc alternator charger, aided by two 120 watt solar modules via a MasterVolt MPPT solar regulator.

The typical daily draw (including losses) is about 900 watt hours (75 Ah/day) of which an average of 750 Wh/day and a peak of 1200 Wh/day is provided by solar. The draw could be reduced by about 20% by replacing the halogen globes with 5 to 8 watt LEDs. As there is plenty of spare roof space it was suggested to increase solar capacity by a further two 120 watt modules. These could be reconnected in series-parallel for 24 volt operation. This enables the existing wiring to be more than adequate - and the MPPT regulator can readily accept that higher voltage. The battery capacity could be increased but as the system now charges well in only partial sun the owner decided not to do so.

A quick 'around-Australia'

A young couple with a few months-old baby decided to do a quick 'around Australia' (of about 26,000 km over 12 months) and bought a secondhand Mercedes Sprinter camper van for this purpose.

With little time for extended stays on-site, extensive energy storage was not needed. A voltage-sensing relay protected the starter battery. The modest electrics (60 litre fridge, 12 volt TV, water pump and a few LED lights) were powered by two alternator-charged 100 Ah Trojan 'traction' batteries. The vehicle had a 15 amp three-stage battery charger, but as the owners stayed in caravan parks every few days to do the baby's washing, etc., it was only needed after stays on remote sites exceeding two nights. Had the couple wished to be free of caravan parks, an EFOY fuel cell would have been useful as back-up. The system worked perfectly throughout. The unit was sold at the end of the trip.

Seven metre caravan (woes)

This caravan, used mainly at weekends, had a 350 Ah battery bank charged from the 4WD's alternator - that fed the caravan via the only too typical 16 metres of (inadequate) twin 4 mm² cable plus a further four conductor metres to the batteries. Charging, at the resultant 13.0 volts, was limited to about 5.0 amps and 50% to 55% charge. The 220 litre electric fridge barely cooled even while driving.

The van had 12 volt incandescent lighting but scarcely enough energy to use it. Fortunately it had a manual water pump and battery radio. Like many with such problems, the owners were asleep by 7.30 pm. Not surprisingly, the batteries needed replacing every 12 months - at over $1000 each time.

An auto electrician's installation of a dc-dc alternator charger in the caravan assisted the fridge, but charging still remained throttled by the existing 4.0 mm² feed cables (that the owners refused to upgrade). The fridge now works as it should while driving but the owners nevertheless complain about its performance on-site despite it being demonstrably due to that absurdly too small cable.

While the helpful auto electrician was blamed, the main failing is incompetent initial design, and the owners curiously illogical unwillingness to install heavier cabling (costing about $50) - let alone supplement the input via solar modules. The work would cost a fraction of ongoing battery replacement.

Fifth wheeler caravans

While this is not a book about caravan dynamic behaviour, the author has an extensive background in vehicle dynamics generally and, as with others in this field, strongly recommends a fifth wheeler if a heavy trailer longer than about six metres (20 ft) is required.

Figure 5.38. Glenn Portch's own 11.2 metre 'Navigator'. Pic: Glenn Portch.

For fifth wheelers, an excellent approach is to have at least 200 watts of solar on the roof of the tow vehicle, plus a 120 Ah battery located on the tray just behind the driving cab - and a solar regulator at the rear of that cab. This can be used to run a permanently-on fridge for shopping, etc.

A second, independent system, with as much solar as needed, is then installed in the fifth wheel caravan, with batteries located low down and close to its centre. If alternator charging is needed, a dc-dc alternator charger, or battery management system, can be installed close to that battery bank. This needs a 13.5 mm² or ideally 16 mm² cable run from that alternator via an Anderson connector.

For an innovative approach to fifth wheeler design, the superb 11.2 metre unit shown here (one of ten built by Glenn Portch) weighs a mere 3100 kg and has 1350 kg payload. The entire cladding is a semi-translucent material that ensures adequate daytime lighting. Its large solar array provides ample power for extended stays on-site. Even the huge fridge is self-built.

Chapter 39

Frequently asked questions

Can I run my 170 litre three-way fridge from solar?

Given enough solar capacity you can run a great deal from an RV's solar but doing so with three-way fridges makes no sense. They draw from 12 to 25 amps or more. Electric-only compressor fridges draw only 1.5-5 amps for the same size and cooling. With three-way fridges use 12 volts only whilst driving, 230 volts when available and otherwise LP gas.

Can I use alternator and solar charging at the same time?

You can parallel inputs if you wish but a second charging source only assists significantly with very large battery banks and/or deeply discharged batteries. Once beyond 50% or so charge, the battery (not the charging source) will limit the rate of charge. Excepting for LiFePO4s, increasing charging capacity will not (above that 50% or so charge) result in batteries charging any faster.

Batteries tend to charge from whichever source provides the highest voltage at any time. This is usually at first from the alternator until 65 to 70% of full-charge, and then from the often higher voltage from a solar regulator when adequate input is available. Recognising this, some battery management systems automatically select whichever source provides the higher input at any time. If you wish simply to parallel the inputs, no precautions are needed. One source will not feed energy into the other, nor overly confuse solar regulation.

What is the best way to maintain an AGM battery when the RV is not in use?

That recommended by many AGM makers is to fully charge, then disconnect the load. Below about 25°C ambient an AGM battery will then maintain at least 60% charge for 12 months or so. Then, again fully charge and so on. Do not attempt to trickle charge as few (if any) chargers can be adjusted to limit current to the truly tiny amount required. AGMs are *known* to be damaged by float charging.

If the RV has solar and is stored where the solar array receives sun, use the excess output for garden lighting or anything similarly useful - such as using the RV's freezer (as we did in Broome) for overload cooling - but not for those batteries.

As my battery bank keeps flattening should I add more batteries?

Chronically flattening batteries is usually caused by too high a load, or insufficient charging capacity. If energy is not coming in, it is not there to store. Adding more batteries can no more assist than opening a second bank account because the first keeps running out of money. You must either increase charging capacity (possibly via solar) and/or reduce the amount of energy you use.

Do I really need to ventilate sealed batteries?

Battery makers warn that sealed batteries have pressure vents that open in the event of a major internal fault - releasing hydrogen that fizzles at about 4% and explodes violently at about 14% concentration, and that adequate ventilation is essential. Follow that advice.

At what voltage is a lead-acid deep-cycle battery fully-charged?

It is impossible to give a meaningful answer except that it should be about 12.7 to 12.8 volts after a conventional deep-cycle battery has rested for at least 24 hours. Any voltage check prior to that is likely to be in error - in that the true voltage may be 0.2 volt or so lower than that shown. That may not seem much but is the difference between (say) 80% and 100% charged. AGMs batteries are likely to reach 12.85 volts at full-charge. Such measurement is likely to be accurate with a few percent after an hour or two with battery totally off-load.

Be aware that a worn-out battery may show 14.4 volts or so across it soon after it is placed on charge. This is a plate surface condition only. That apparent charge will remain there for only a few minutes after any load is applied.

A perfectly good well charged deep-cycle battery may show as little as 11 volts for some time after running a microwave oven. This too is *totally* normal.

Chapter 40

Import electrical issues

In late 2009, the Australian Federal Infrastructure, Transport, Regional Development and Local Government Minister, Anthony Albanese, stated that "importers and local manufacturers could face large fines (of up to $66,000) under new safety standards that come into force today" [November 20, 2009]. "The new rules apply to all imported and locally-produced trailers up to 4.5 tonne - including caravans and other recreational trailers." The minister added: "this move will still allow Australians returning home from long periods overseas to bring with them their personal vehicles, while reducing the potential for abuse by organised syndicates seeking to bypass our rigorous motor vehicle certification arrangements by using third parties to import non-compliant vehicles".

While it covers ADR requirements, a loophole (intended to enable overseas visitors to use 110 volt 60 Hz razors, phone chargers, etc.,) still allows non-electrically compliant RVs bought overseas to be imported and used by their original owner (by adding a 230-120 volt transformer) but does *not* make it fully electrically compliant. It must not be sold, or even given away, unless brought to full and re-certified compliance. As this necessitates replacing all cabling and electrical fittings, it is a major task involving, in some cases, the entire outer cladding being removed and replaced.

The main safety issue

The main safety issue involved is the use of a 230-120 volt transformer. These are intended to power only one Class A appliance (i.e. that has its metal chassis or enclosure connected to earth). Electricians confirm there are safety issues if more than one such appliance is connected. AS/NZS 3000:2018 requires that 'all live parts of a separated circuit shall be reliably and effectively separated from all other circuits, including other separated circuits and earth'.

Even if such transformer usage was safe, all electrical goods sold in Australia must meet the requirements of the Electricity (Consumer Safety) Act 2004 and its regulations - that 'Declared Items' must be officially approved and certified. The listing is state administered but that for NSW is typical. It includes everything electrical: from supply cords, plugs and socket outlets, to electric blankets, fridges, etc. It precludes *any* 110 volt, 60 Hz item. That list can be accessed at: http://www.fairtrading.nsw.gov.au/pdfs/Businesses/Explanatory_notes_and_declaring_order.pdf. There are similar lists in other states.

This section of the book has been prepared with the assistance of a highly experienced licensed electrician with Standards Australia connections, and professional electrical engineers. All state the existing practice is unsafe, e.g. "there is no way an isolation transformer either permanently wired or portable and plugged into the existing van power inlet can comply with AS/NZS Standards, things are either dangerous or they are not", says one. Several electrical engineers involved with Standards Australia, noted that body had neither foreseen nor intended the Standards to be interpreted this way.

The seemingly inescapable conclusion is that there has been a disconnect between authorities. Regardless of 'selling considerations' that may apply to electric razors and such like, trade regulation considerations cannot realistically be extended to imported RV electrical systems and appliances. Such units are electrically unsafe unless brought into full compliance. And that totally excludes 120 volt 60 Hz appliances. These are Declared Items under the 2004 Act.

As reselling anything with an non-electrically compliant system is a serious offence in many jurisdictions, and a criminal offence in at least one, intending buyers/sellers are advised to seek expert legal advice. Few original owners of these imports have proved to be aware of this situation: most believe their vehicles are 100% compliant. They are not.

Chapter 41

Contacts & references

Early editions of this book included supplier contact details that quickly became outdated. The ease of locating via Google and the internet resolved this need. Now, many companies wish to use email only and deliberately omit their postal address in their promotions. There is also a proliferation of local and overseas on-line vendors, and many existing vendors now solicit only internet sales.

'Googling' what you seek will return many links, some of which may not be current or exactly what you are looking for, but is generally superior to most printed lists. For these and other reasons this book no longer lists contact details in printed form of anything likely to change.

Companies can often be located via their trade name followed by .com.au (for some Australian companies). Others have only .com. There are also geographical and organisational suffixes: e.g. New Zealand uses .nz. Organisations are likely to use .org.

Internet Explorer originally required you to enter the prefix www. This is no longer needed and it may bother some systems if used.

Where the company name is more than one word try joining the words together, omitting upper case letters, and spelling out '&' as 'and'. Where the above does not work try using a hyphen: e.g: eastpenn-deka.com, or an underline, eastpenn_deka.

Our all-new main website (rvbooks.com.au) contains some of my previously published articles over a wide range of topics and is constantly expanded and updated. It also has Links to other sites of interest.

Some useful information on electrical systems can be found on web forums but most is worthless and sometimes downright dangerous were it to be followed. Forums are occasionally handy if you know already know a fair amount about the topic, but rarely otherwise.

The main references in this area (mostly for mains-voltage systems), applicable for both Australia and New Zealand, are the primary Standard - known also The Wiring Rules - now AS/NZS 3000.2018, and the RV-related AS/NZS 3001.2008 (as Amended in 2012). They are very different from the earlier versions. They are costly and unfortunately can no longer (at present) be borrowed from public libraries.

Because of the current requirement for all except extra-low voltage installation to be done by licensed electricians (and even that is now changing), there is little else currently written (except at text book level) on mains-voltage installation in Australia. There is also little on 12/24 volt wiring excepting for this book, but its use in connection with solar is covered in depth in Collyn Rivers' books *Solar Success* (for homes and properties and *Solar That Really Works!* (for boats, cabins and RVs). See also our solarbooks.com.au

Caravan & Motorhome Electrics

rvbooks.com.au

Table of Contents

Chapter 1 - Terminology	3
Electrical units & terms	3
Chapter 2 - Basic Electrics	6
Electron flow	8
AC/DC explained	8
Voltage, current & resistance	8
Energy & power	9
Power factor	9
Chapter 3 - Overview of an RV's electrical needs - The early days	11
Improved batteries	13
Lighting	14
Appliances	14
CPAP	15
Refrigeration	16
Water pumps	16
Air conditioning	16
Converters	16
Imported RVs	17
Chapter 4 - Providing the power	18
Caravan park power	18
Away from mains power	18
Staying longer on-site	20
Recommended approach to battery and solar required	20
Battery capacity	21
Solar capacity	22
Other sources of power	22
Chapter 5 - Scaling the power required	23
Table 1 - typical energy draw.	23
Energy draw of proposed lights and appliances	24
Table 2 - proposed energy draw	25
Chapter 6 - Installing safety and legality	26
Australia/NZ regulations	27
Chapter 7 - Installing 12/24 volt wiring	28
Current ratings	29
AWG/B&S	29
Cable size needed	29
Table 3: voltage drop formula	30
Earth return	31
Table 4: Cable sizes compared	31
Tracking voltage drop	32
Cable protection	33
Circuit breakers	33
Fuses	34
Winch solenoids	36
Voltage-sensing relays	36
Variable voltage alternators	36
Plugs & sockets	37
Extra-low voltage switches	37
Switch & meter panels	38
Current shunts	38
Crimp connectors	40
Correct crimping	40
Power posts/connector boxes	41
The wiring layout	42
Chapter 8 - Mains-voltage wiring	44
How electrical earthing works	45
Functional earthing	46
RCD/CB protection	46
Cable Rating	46
Earthing is still essential.	47
Never join supply cables together	47
Caravan park supplies	47
Ten to 15 amp cable issues	47
Polarity explained	48
Polarity checking	49
Double-pole switching	50
Cable sizing & installation	50
Inverters	50
Generators	51
Change-over switches	52
Lights & appliances	52
Disclaimer (professional)	52
Chapter 9 - Batteries (general)	54
Construction of a lead-acid battery	54
Starter batteries	55
Cold-cranking amps (CCA)	57
Reserve Capacity	57
Lead/calcium batteries	57
Deep-cycle batteries	58
Marine batteries	59
Sealed batteries	59
Gel cell batteries	59
Absorbed glass mat batteries	60
Lithium batteries	61
Battery charge & life	62
Capacity - how much?	63
Self-discharge	63
What type of battery to choose	63
	64

Chapter 10 - Installing batteries	
Ventilation is still essential	64
How much ventilation needed	65
Series or parallel?	65
Which is better?	66
12 volts from 24 volts	66
Voltage equalisers	67
Voltage Converters	67
Chapter 11 - Battery charging (general)	69
Multiple power point tracking (MPPT)	70
Multi-stage charging	70
Overcharging	72
Undercharging/over-discharging	72
Charging efficiency	72
Battery capacity & temperature	73
Self-discharge	73
Pulsing	74
Assessing battery condition	74
Lithium	74
Mains voltage battery chargers	75
RV Electrical Converters	75
Introduced voltage drop	77
Chapter 12 - Battery charging (alternators)	78
Alternator voltage regulator	79
Ability to engine restart	80
Voltage-sensed switching	81
Emergency starting	81
Regenerative braking	82
The need for change	83
The CAN bus	83
Chapter 13 - de-de alternator charging	84
Chapter 14 - Installing a de-de alternator charger	85
Chapter 15 - Variable voltage alternator charging	86
Identifying alternators	87
Chapter 16 - Inverters	88
Transformer-based	88
Switch-mode	88
Modified square-wave inverters	89
Freestanding inverters	89
Wired in	90
Inverter/chargers	90
Phantom loads	91
Microwave ovens in RVs	92
Assessing inverter size	92
Chapter 17 - Installing an inverter	93
Connecting an inverter	93
Chapter 18 - Generators	95
What appliances draw	95
Sizing the generator	96
Inbuilt motor-generators	96
The '12-volt de output'	97
Twelve/twenty four volt generators	97
Building your own	98
Chapter 19 - Installing a generator	99
Stalling on st art -up (issues with)	99
Quietening noisy generators	100
Small generator problems	101
Chapter 20 - Wind power generators	103
Wind run	104
Fishing camps	104
Propeller types	105
Propeller braking	106
Evaluating & buying a wind generator	106
Free power (myth of)	107
Installation - do what the makers say	107
Chapter 21 - Fuel cells	108
Brief fuel cell history	108
How fuel cells work	109
Fuel cell output	109
LP gas fuel cell	110
Fuel cell efficiency	111
Fuel cell future	111
Chapter 22 - Solar energy	113
Assessing solar output	113
Solar up north	114
Solar module types	115
Monocrystalline	115
Polycrystalline	115
Amorphous	116
Solar module reality	117
Energy mismatch	118
Fixed & loose solar modules	118
Dual systems	119
Scaling batteries to charging output	120
Chapter 23 - Installing solar modules	122
Leave an air space	122
Voltage/current	123
Measuring module output	124
Solar measuring via multimeter	124
Solar measuring via clamp amp meter	124
Solar module voltage measure-	125

ment	
Solar module care	125
Solar module protection	126
Solar module life	126
Chapter 24 - Solar regulators	127
Buying a regulator	127
Solar regulators for wind power generators	128
Programming a regulator	128
Chapter 25 - Installing solar regulators	129
Monitoring issues	130
Chapter 26 - Energy monitoring	131
Chapter 27 - Installing energy monitors	133
Current shunts	133
Exceptional connection	135
Chapter 28 - Lighting	136
Incandescent lights	136
Halogen	136
Fluorescent lights	137
LEDs	137
LED light output	139
Installation	139
Chapter 29 - Water	140
Controlling the flow	141
Accumulator tanks (pressure vessels)	141
Filters	143
Chapter 30 - Electric toilets	144
Macerator systems	144
Chapter 31 - Refrigerators	146
Absorption (three way)	146
Compression type	147
Assessing energy usage	149
Top versus front opening	150
Freezers	151
Fridge standards	151
The end result	151
Chapter 32 - Installing & optimising fridges	153
Reducing heat loss	156
Domestic fridges in RVs	156
Building your own fridge	157
Chapter 33 - Television	159
VHF/UHF antennas	160
Connecting the antenna	160
Antenna amplifiers	160
Satellite television	160
Television receivers	162
Standby consumption	163

Chapter 34 - Communications	164
Satellite communications	165
High frequency (HF) radio	166
UHF CB radio	166
Useful links	167
Chapter 35 - Electrical & radio interference	168
Chapter 36 - Lightning protection	170
Chapter 37 - Caravan specific issues	172
Delusions of adequacy	172
Trailer connectors - Australia	173
Pin numbering	173
Table 11: 7/12 pin connections	174
Trailer connectors (13-pin) - Europe	175
Table 12: 13 pin connections	175
Electric brakes	176
Electric brake connection	177
Braking problems	177
Random brake operation	178
AL-KO ESC	178
Tuson (Dexter) ESC	178
Chapter 38 - Example systems	179
Typical basic usage	179
Fully off-road (OKA)	179
Nissan Patrol/TVan	181
A 24 volt system	182
A quick 'around-Australia'	182
Seven metre caravan (wres)	182
Fifth wheeler caravans	183
Chapter 39 - Frequently asked questions	185
Chapter 40 - Import electrical issues	187
The main safety issue	187
Chapter 41 - Contacts & references	189

www.ingramcontent.com/pod-product-compliance
Lightning Source LLC
Chambersburg PA
CBHW071919290426
44110CB00013B/1411